A Cloud a Day

A Cloud a Day

GAVIN PRETOR-PINNEY

CHRONICLE BOOKS
SAN FRANCISCO

This book is dedicated to all the members of the Cloud Appreciation Society.

First published in the United States of America in 2019 by Chronicle Books LLC.
Originally published in the United Kingdom in 2019 by Batsford.

Library of Congress Cataloging-in-Publication Data available.

ISBN 978-1-4521-8096-0

Manufactured in China.

Design by Jon Glick.
Cover image © William Hawkins.

PAGE 2 Fluctus wave formations, also known as Kelvin-Helmholtz clouds, forming in fog spotted by Pat Cooper over Bridgnorth, Shropshire, England.

10 9 8 7 6 5 4 3 2

Chronicle books and gifts are available at special quantity discounts to corporations, professional associations, literacy programs, and other organizations. For details and discount information, please contact our premiums department at corporatesales@chroniclebooks.com or at 1-800-759-0190.

Chronicle Books LLC
680 Second Street
San Francisco, California 94107
www.chroniclebooks.com

CONTENTS

THE CLOUD TYPES

SELECTED HIGHLIGHTS

THE 10 MAIN TYPES KNOWN AS GENERA

SPECIES AND VARIETIES

OTHER CLOUDS

SUPPLEMENTARY FEATURES AND ACCESSORY CLOUDS

INTRODUCTION

IT IS EASY TO FORGET that you live in the sky—not beneath it, but within it. Our a tmosphere is an enormous ocean, and you inhabit it. This ocean is made up of the gases of air rather than liquid water, but it is as much of an ocean as the Atlantic or the Pacific. You may think of yourself as living on the ground, but all that means is that you are a creature of the ocean bed. You still inhabit the atmosphere like a sea creature does the water.

"It is a strange thing how little in general people know about the sky," wrote the Victorian art critic John Ruskin (see page 40). Strange indeed, given how important it is to us. One reason for this might be that the sky is always there. It is the ever-present backdrop to our lives, and anything as ubiquitous as this is easily missed because it hides in plain sight.

We at the Cloud Appreciation Society believe that you would do well to pay more attention to the sky. Having your head in the clouds, even for just a few moments each day, is good for your mind, good for your body, and good for your soul. This book aims to show you why.

"The first step to wisdom," as the biologist E. O. Wilson noted (see page 161), "is getting things by their right names." Learning the names for a few of the different cloud types is a good way to start a new relationship with the sky. Every cloud is unique, but we humans love to put things into groups and so we gather their chaotic forms according to ten main types, known as cloud genera. You might have learned some of them at school—names like Cumulus, Stratus, and Cirrus. There are also many subcategories of cloud. These cloud species and varieties and cloud features crop up here and there among the main types. Some of them are rare and fleeting, and you have to really pay attention to the sky to be able to spot them. To start getting used to which cloud is which, you can navigate your way through the notable examples using the **Cloud Types map**.

CLOUDS THAT LOOK LIKE THINGS

THINGS

ANIMALS

PEOPLE

The Latin names sound formal, but they are mostly just based on how the clouds look—on their shapes—and you certainly don't need to remember Latin terms to enjoy finding shapes in the clouds. You might remember doing this when you were young. Back when you had time on your hands, when the only deadline was bedtime, and your imagination could float free.

Finding shapes in the clouds is how most of us first become interested in the sky. There is an aimless pleasure to this side of cloudspotting, one that feels nostalgic. The early age at which this relationship with clouds is first forged might explain why our feelings about clouds and the sky run deep. But once we have grown up, this aimless pleasure of youth feels frivolous. We've no time to sit around gazing at clouds.

So why when you speak to someone who, for whatever reason, knows their days on Earth are numbered, do you often find them saying that the sky, the transient, ephemeral, ever-changing clouds, feel more worthy of their attentions than most of the stuff we spend our days worrying about? Just because something is aimless doesn't mean that it is pointless.

So find the time, every now and then, "to make the shifting clouds be what you please," as the poet Samuel Taylor Coleridge put it (see page 214). You will be engaging the idle mode of your brain—a mode that has been effectively eradicated from our daily lives by device culture—because cloudspotting legitimizes doing nothing. You'll be freeing your imagination and reminding yourself to stay lighthearted. The spottings in this book by members of the Cloud Appreciation Society will help get you in the mood. Use the **Clouds That Look Like Things** map to find them.

We all know clouds can have a profound effect on our moods. No wonder, then, that they are the tool artists use to introduce feeling to a landscape painting. The 19th-century Romantic painter John Constable argued (see page 32) that in any landscape painting the sky is the "chief organ of sentiment," the "key note." You can map our changing attitudes towards nature through a history of how we depict the sky in art. Until the last 200 years or so, the sky was treated in Western art less as the key note and more as a footnote. When clouds did appear, they were little more than background decoration, compositional space-fillers, at best cartoon cushions for deities to recline on. They rarely played a major role. But there were exceptions, and you can navigate these alongside more modern artistic explorations of the atmosphere with the **Art of the Sky map.**

ART OF THE SKY

BEFORE 1300S

13TH–18TH CENTURIES

19TH CENTURY

20TH CENTURY

21ST CENTURY

To tune into the sky is to slow down. Clouds may be in a state of perpetual change, but it is one that, more often than not, appears gradual. They might in fact be moving quite fast—ice crystals in the high, sweeping streaks known as Cirrus can blow along at speeds approaching 300 kilometers (200 miles) per hour—but they appear to change gradually because they are a long way away. So you can treat cloudspotting like a moment of meditation, a meditation on the sky, which differs from other forms in one important regard: What you are concentrating on, the sky, is beyond your control. You cannot plan your cloudspotting for a specific duration, schedule it for a certain moment in the day. Cloudspotting is a frame of mind more than a planned activity. When the sky puts on a show, you just have to be prepared to pause what you are doing and engage with it for a few moments.

Clouds are the embodiment of chaos and complexity. Why do they change so unpredictably? What accounts for their dynamic, ever-shifting forms? The answer is a simple one. Clouds change appearance so much because of the unique qualities of water. It is the only substance on our planet that is found naturally in the three states of solid, liquid, and gas. Nothing else on Earth (or above it) shifts between these states with such ease, ushered along by just the slightest changes in temperature. And one of the three, the gas state of water, is invisible. It is known as water vapor, and it is transparent. So a subtle warming or cooling of the air can be enough to choreograph the magical dance of water from invisible to visible, from transparent gas to a solid-looking array of droplets or a translucent streak of ice crystals.

We see clouds because their particles reflect and scatter the sunlight. And sometimes when they do so, if their droplets or ice crystals are just the right sizes, shapes, and orientations, they can bend and separate the light to form a whole range of arcs, rings, bands, and spots of color. Journey through these light phenomena of the atmosphere with the help of the **Optical Effects map**.

Each of the 365 clouds in this book, no matter whether it was spotted by an astronaut aboard the International Space Station, an old master of the Dutch Golden Age, or a member of the Cloud Appreciation Society from their backyard, should be thought of as a reminder. Each cloud is a tap on your shoulder, prompting you to look up, take a breath, and unfetter your thoughts from earthbound concerns. It is there to remind you to look around you, look above you, and appreciate this ever-changing ocean of air that we all inhabit and share.

OPTICAL EFFECTS

CAUSED BY DROPLETS

CAUSED BY ICE CRYSTALS

CAUSED BY EITHER DROPLETS OR ICE CRYSTALS

ELECTRICAL

ACKNOWLEDGMENTS

I am very grateful to all the members and friends of the Cloud Appreciation Society who contributed images to the book. I would personally like to thank the following members whose words and ideas featured in the Cloud-a-Days: Yoav Daniel Bar-Ness (Member 10,389), Sheila Brooke (Member 32,250), Shelley Collins (Member 9,733), Elliot Davies (Member 7,143), Kym Druitt (Member 19,908), William A. Edmundson (Member 5,218), Jeanne Hatfield (Member 36,420), Richard Joosse (Member 32,314), Andrew Pothecary (Member 3,769), and Judith Strawser (Member 32,075). In particular, Elliot Chandler (Member 16, 353) has been instrumental in helping write and develop much of the content, for which I am especially grateful.

—Gavin Pretor-Pinney

THE CLOUDS

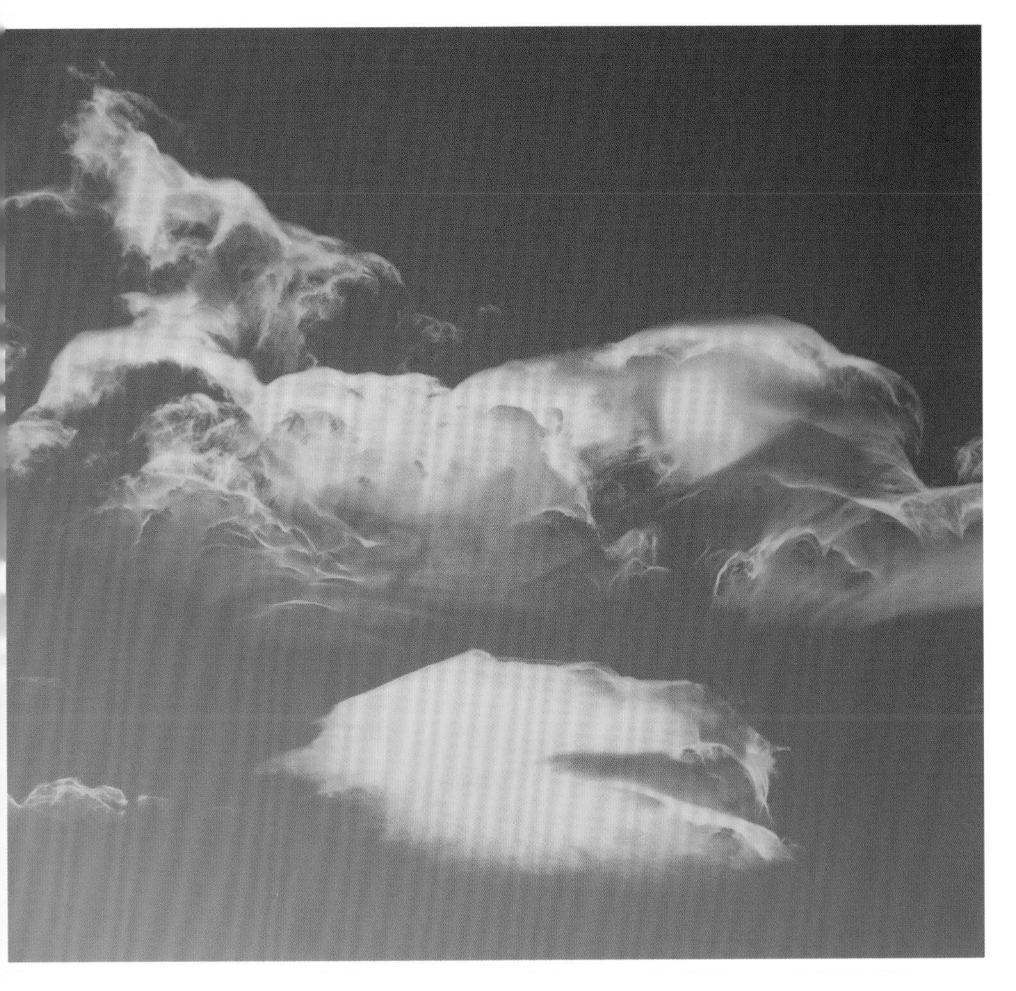

One way to open your eyes to unnoticed beauty is to ask yourself, "What if I had never seen this before? What if I knew I would never see it again?"

From *The Sense of Wonder* (1956) by Rachel Carson.

ABOVE Cirrocumulus lenticularis clouds contorted by turbulence in the lee of the Sierra Nevada mountains, California, US, spotted by Stephen Ingram (Member 7,328).

WHEN DID REALISTIC CLOUDS first appear in pictures? It was most likely in 12th-century Chinese art. This detail shows Stratus fractus clouds on the slopes of mountains in an ink drawing by the Chinese artist Mi Youren, created sometime before the year 1200. In another of his several *Cloudy Mountains* drawings the artist included an inscription:

Innumerable are the wonderful mountain peaks which join the end of the sky,

Clear or cloudy, day or night, the misty atmosphere is lovely.

THE LENTICULARIS CLOUD, which has a distinctive disc shape and forms in the vicinity of mountains, is named after the Latin word for "lentil." This is because no one could think of the Latin for a flying saucer.

ABOVE Altocumulus lenticularis, spotted by Ian Boyd Young over the east end of the French Pyrenees mountain range, near the border with Spain.

OPPOSITE *Cloudy Mountains* (before 1200) by Mi Youren. Handscroll, ink on paper.

BANDS OF PASTEL COLORS sometimes appear spread across high clouds like this Cirrostratus. They're caused by the sunlight bending as it passes around the tiny particles of the cloud. The optical effect is known as diffraction, and the different wavelengths that make up sunlight bend by different amounts. This has the effect of separating out the wavelengths, which in isolation appear colored. The colors only show when a cloud's droplets or ice crystals have a very consistent size and are very small—each one around a thousandth of the width of a human hair.

ABOVE Cloud iridescence appearing in Cirrostratus, spotted above Cumulus over North Elmham, Norfolk, England, by Danielle Malone (Member 35,276).

It was a hard thing to undo this knot. / The rainbow shines, but only in the thought / Of him that looks. Yet not in that alone, / For who makes rainbows by invention? / And many standing round a waterfall / See one bow each, yet not the same to all, / But each a hand's breadth further than the next. / The sun on falling waters writes the text / Which yet is in the eye or in the thought. / It was a hard thing to undo this knot.

Unfinished poem, dated 1864, by Gerard Manley Hopkins.

ABOVE A rainbow spotted in the spray by Sarah Jameson, standing around a waterfall in Skogar, Southern Region, Iceland.

FLUCTUS CLOUDS are also known as Kelvin-Helmholtz wave clouds, after 19th-century scientists Lord Kelvin and Hermann von Helmholtz, who studied turbulence at the boundary between moving fluids. A difference in velocity between two airstreams, known as wind shear, results in unstable vortices that can lead to this rare and fleeting formation.

ABOVE Fluctus clouds curling like peaks of meringue over the Green Mountains near Starksboro, Vermont, US, spotted by Keith Edmunds (Member 41,937).

A PILEUS WAS A BRIMLESS FELT HAT worn by the Ancient Greeks and later the Romans. It is also the name of this smooth cap of cloud that can appear over the top of a towering Cumulus as it builds rapidly upwards through the atmosphere. The formation is caused by airstreams passing above being cooled as they are lifted by the violent rising air currents within the convection cloud below. For Romans, the *pileus* was a symbol of freedom. It was given to slaves when they were released from servitude.

ABOVE Pileus forming above towering Cumulus clouds, spotted over Kentucky, US, by Frank Leferink (Member 41,121).

ABOVE A cat stalks headlights in Ashtead, Surrey, England. Luckily, the alarm has been raised as this cat wears a bell. Spotted by Debbie Whatt (Member 43,013) and known also as Cumulonimbus capillatus.

OPPOSITE Cirrus uncinus, spotted over Gevninge, Zealand, Denmark, by Søren Hauge (Member 33,981).

The cloud-capp'd tow'rs, the gorgeous palaces,
The solemn temples, the great globe itself,
Yea, all which it inherit, shall dissolve,
And, like this insubstantial pageant faded,
Leave not a rack behind. We are such stuff
As dreams are made on; and our little life
Is rounded with a sleep.

From *The Tempest* (1623), Act IV, Scene 1, by William Shakespeare.

ON SATURN, A RING OF CLOUDS EXISTS with a distinctly angular form that has not changed for 30 years. This hexagon of clouds at the planet's north pole is so large that each side is about the diameter of Earth. The formation was first observed in the early 1980s. While it is understood that the shape results from the path of a jet stream, there is yet to be any clear consensus on what causes it to have such a regular pattern. Whatever the explanation, Saturn's hexagon is a rare example of order emerging from the chaos of clouds.

ABOVE The cloud pattern known as Saturn's Hexagon, spotted in 2016 over the planet's north pole by NASA's Cassini spacecraft.

THIS PAINTING, CALLED *NEW MEXICO RECOLLECTION #12*, is by the American modernist painter Marsden Hartley. The artist produced it in the early 1920s, when he was living in Berlin. It is one of a series based on his memories of the arid landscape of New Mexico, which he described as "essentially a sculptural country." Above the stylized terrain, Hartley depicted what appear to be stacked lenticularis clouds. This is a formation that looks like discs of cloud, layered one on top of another. Such a cloud type would have made complete sense as it is one of the most memorable and, without a doubt, the most sculptural of all the cloud formations.

ABOVE *New Mexico Recollection #12* (1922–1923) by Marsden Hartley. Spotted by Minnie Biggs (Member 4,330).

THE COLOR OF CLOUDS depends greatly on the color of the direct sunlight striking them. The color of sunlight depends greatly on the journey it has traveled through our atmosphere, which scatters out the blue end of its spectrum more than the red. Here, the lower Cumulus and Stratocumulus clouds look pink, while the high tops of the Cumulonimbus clouds look white. The sunrise lit the low clouds by shining at a shallow angle through the denser, lower atmosphere. Its bluer wavelengths having been scattered away along the journey, the light reached the low clouds with peachy, salmon hues. The light shining onto the towering summits of Cumulonimbus traveled through the less dense, higher atmosphere. Its spectrum remained largely intact, and it lit the lofty cloud peaks in pure, brilliant white.

OPPOSITE Clouds at sunrise over the coast of Guatemala, as viewed from directly above by astronauts aboard the International Space Station.

I look about; and should the guide I choose /
Be nothing better than a wandering cloud /
I cannot miss my way.

From *The Prelude* (1850), Book One,
by William Wordsworth.

ABOVE A wandering Cumulus, spotted by Rauwerd Roosen over the Soiern Mountains, Upper Bavaria, Germany.

OPPOSITE A Von Kármán vortex street in marine Cumulus downwind of Isla Socorro, 500 kilometers (just over 300 miles) off the Pacific coast of Mexico, spotted by NASA's Aqua satellite.

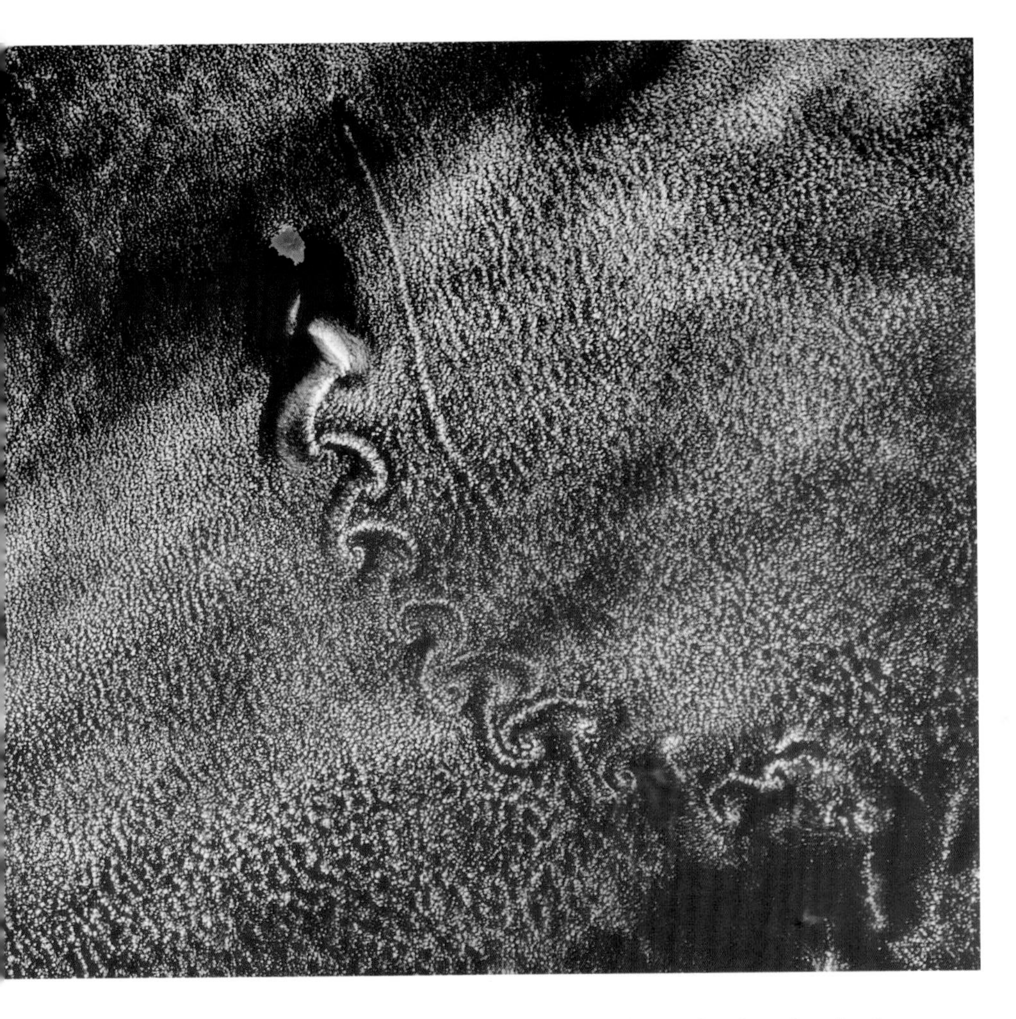

THIS SWIRLING PATTERN appeared in the marine Cumulus clouds downwind of Isla Socorro, a volcanic island in the Pacific Ocean. It is known as a Von Kármán vortex street, named after the Hungarian-American mathematician and physicist Theodore von Kármán, who explained in the 1910s how alternating oscillations can develop as a fluid flows around a blunt obstacle. This is the phenomenon that causes overhead cables to hum a note in the wind. And it is how a solitary volcano jutting up from the ocean's surface can send the clouds into a spin.

SOMETIMES A SHOWER just doesn't cover enough of the sky to form a proper rainbow. The color in this evening downpour might better be described as a "rainsquare."

ABOVE A fragment of a rainbow, spotted in a shower over the Sierra Almijara mountains, Andalucia, Spain, by Rodney Jones (Member 15,695).

OPPOSITE Altocumulus stratiformis perlucidus, spotted over Dorset, England, by Poppy Jenkinson (Member 39,335).

HERE IS HOW THE NAME FOR A CLOUD like this Altocumulus stratiformis perlucidus is constructed. “Altocumulus” is the genus. A genus is one of ten main types into which most clouds can be classified. The Altocumulus genus refers to a clumpy cloud, up at the mid-level of the troposphere. “Stratiformis” is the species. It means that the layer of clumps extends over a large region of the sky. “Perlucidus” is the variety. It refers to when the clumps have gaps between them, rather than being joined into a more continuous layer. In other words, it means “those nice little puffy ones that spread across the sky,” but in Latin, to make it sound official.

FROM 1820 TO 1822, the English Romantic landscape artist John Constable dispensed with painting the land altogether and concentrated solely on the sky. During this period, he produced numerous Cloud Studies—fast sketches of the sky, never intended for exhibition. These small canvases, far more loose and free than his famous landscapes, were Constable's experiments in depicting the moods of the sky. Through this "skying," as he called it, Constable strove to become a master of clouds. As he wrote in a letter to his friend John Fisher in 1821, Constable considered the sky to be "the chief Organ of Sentiment" in any landscape painting—the "key note," which brings emotion and drama to the scene.

RADIATUS CLOUDS ARE WHERE PARALLEL LINES or rolls extend so far across the sky that they appear to fan out from a point on the horizon. This is just an aerial version of that perspective effect in which train tracks appear to converge in the distance when you look down the length of them.

ABOVE Altocumulus radiatus, spotted over Prospect Park, New York, US, by Elise Bloustein (Member 41,703).

OPPOSITE Eight of the 14 Cloud Studies by John Constable within the collection of the Yale Center for British Art, New Haven, Connecticut, US, spotted by Mark Richardson (Member 42,827).

THESE POUCH-LIKE LOBES of cloud are known as mamma, from the Latin for "udder" or "breast." The exact mechanism for their formation is not fully understood. It might relate to air cooling within the cloud layer above—perhaps as ice crystals in the cloud melt. As it cooled, the air would become denser and sink below, arranging itself into pockets of descending air.

A GLORY IS AN OPTICAL EFFECT that can appear when sunlight shines from behind the viewer onto a layer of cloud. Colored rings appear around the observer's shadow due to the sunlight being diffracted as it is reflected by the droplets of the cloud. The effect is also known as a Brocken spectre by mountain climbers, who sometimes see it as they climb above a cloud layer. The most common place to see a glory these days is from the window of an aircraft. Look for it when your plane's shadow is cast down onto a cloud layer below. The laws of optics decree that the circle of colors is centered on your position within the aircraft. For this reason, you can tell from a photograph of an aircraft glory where the cloudspotter was sitting—here, just behind the wing in the top image, and in the cockpit, flying the plane, in the lower image.

TOP Spotted by Ram Broekaert over Enghien, Hainaut Province, Belgium. **BOTTOM** Spotted from the aircraft cockpit on an approach to Geneva Airport by Richard Ghorbal (Member 5,117).

OPPOSITE Mamma clouds, spotted over Cap Ferret, Aquitaine, France, by Katalin Vancsura (Member 30,830).

THIS STORM SYSTEM came barreling in over the coast of Queensland, Australia. The upper left part of the image shows the spreading canopy up at the top of the Cumulonimbus storm cloud. The low edge of cloud, looking like the skirt of a hovercraft, is a feature called arcus, which is also known as a shelf cloud. This is where cool air dragged down to the land or sea surface by all the storm's precipitation, visible off in the distance, splays outwards and burrows beneath the warmer, moister, less dense air ahead of the storm. As this lifts, it forms a low, menacing base extending from the front of the storm, heralding the downpours to come.

ABOVE An arcus cloud feature at the base of a Cumulonimbus, spotted by Ebony Willson approaching Moreton Bay, Queensland, Australia.

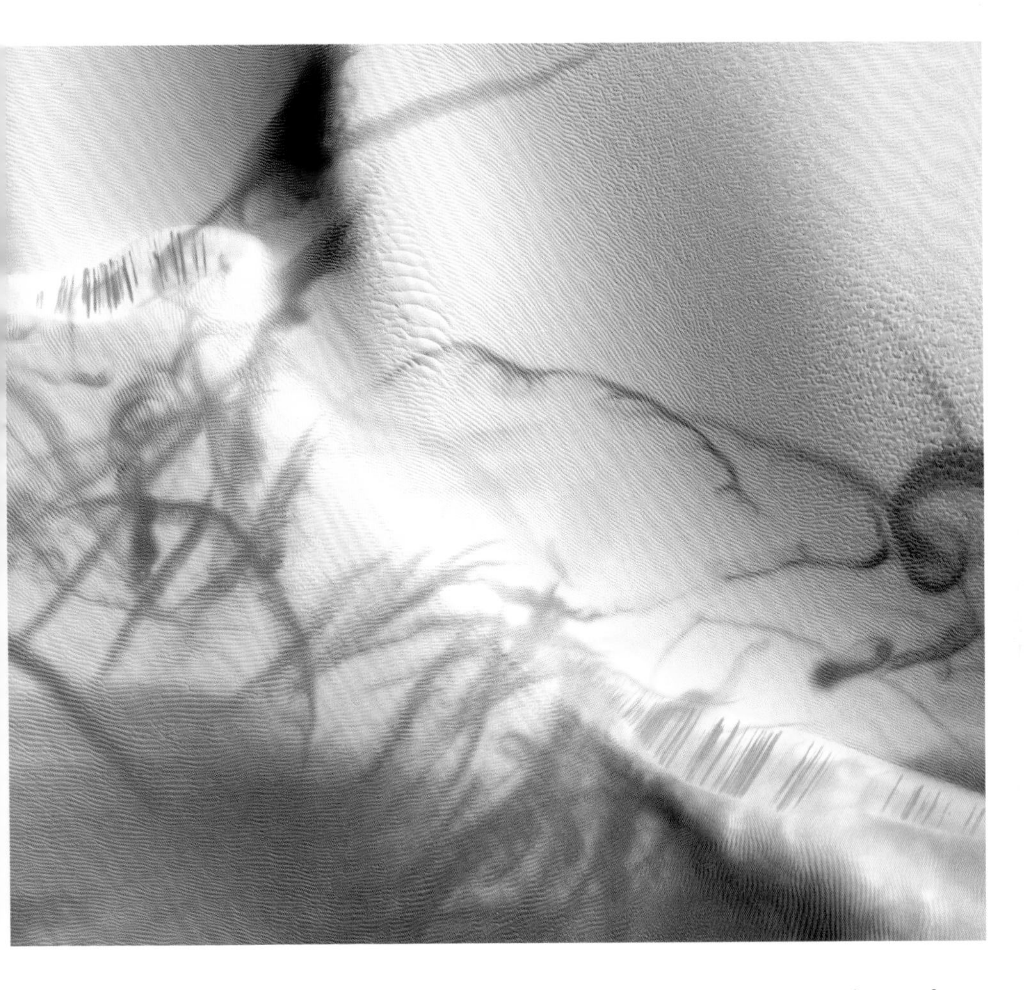

NO, THIS ISN'T A CLOSE-UP of someone's belly tattoos. It is an image of the meandering trails inked on the surface of Mars by dust devils. Growing kilometers high and meters wide, these Martian whirlwinds dwarf their terrestrial cousins. They spin dust and sand at such velocities that they produce miniature lightning bolts. One such "Martian devil" did such a good job of cleaning the solar panels on Spirit, the Mars Exploration Rover, that power levels dramatically increased after the encounter.

ABOVE Dust devil tracks spotted on the surface of Mars by the HiRISE camera on board NASA's Mars Reconnaissance Orbiter.

WHEN THE ASPERITAS CLOUD was accepted as a new classification in 2017 by the World Meteorological Organization (WMO), everyone asked if the formation itself was new or just the name. When we first proposed it as a new classification, we argued that the formation, though rare, had always been around. Smartphones just meant anyone could capture a cloud and send it in to us, so we had spotted a pattern previously missed. German artist Caspar David Friedrich spotted an asperitas back in 1835. With no smartphone to hand, he'd depicted it in his painting *Seashore by Moonlight*. We like to think that had the Society existed in 1835 Friedrich too would have sent his asperitas spotting in to us. Would be worth a lot of money for the Society today. Unless, of course, we'd been dumb enough to forward it on to the WMO.

ABOVE Asperitas clouds depicted in *Seashore by Moonlight* (c. 1835–36) by Caspar David Friedrich.

OPPOSITE Nacreous clouds, spotted over The Curragh, County Kildare, Ireland, by Kelly Hamilton (Member 15,098).

MID-WINTER IS THE TIME OF YEAR to spot nacreous clouds. Also known as polar stratospheric clouds, these form higher than the normal weather clouds, at altitudes of 10–25 kilometers (9–16 miles), which is inside the stratosphere. They are generally only observed from latitudes of more than 50 degrees. Being so high, nacreous clouds can catch the light when the Sun is below the horizon, and they look brightest against the darkened sky within an hour or two before sunrise or after sunset. This is when the lower part of the sky (the troposphere) is in shadow. The cloud's particles, which can consist of different combinations of acid and water, diffract the sunlight and separate it into beautiful bands of pastel colors. No wonder these formations also go by the name of "mother-of-pearl clouds."

It is a strange thing how little in general people know about the sky. It is the part of creation in which nature has done more for the sake of pleasing man, more, for the sole and evident purpose of talking to him and teaching him, than in any other of her works, and it is just the part in which we least attend to her.

From *Modern Painters* (1843), Volume I, Section III, by John Ruskin.

ABOVE Cirrus uncinus, spotted by Sarah Nicholson over Kommetjie, Cape Town, South Africa.

OPPOSITE Enhanced-color view of Jupiter's cloud tops processed by Bjorn Jonsson using data from the JunoCam instrument on NASA's Juno spacecraft.

THE GIANT PLANET JUPITER has the largest atmosphere in the solar system. It is perpetually covered with clouds of ammonia crystals. There may also be a thin layer of water clouds hidden beneath. The clouds are arranged into bands of lighter-hued “zones” and darker “belts,” which rotate around the planet in opposing directions at speeds of around 360 kilometers per hour (220 miles per hour). The interaction between these conflicting circulation patterns results in turbulence, which can develop into massive storm systems. Jupiter’s most famous storm is known as the Great Red Spot, but the planet also has many lesser, unnamed storms. These appear as brown or white ovals within the ever-shifting swirls and eddies of cloud. Some Jovian storms last a few hours; others persist for centuries.

The fog comes
on little cat feet.

It sits looking
over harbor and city
on silent haunches
and then moves on.

"Fog," *Chicago Poems* (1916), by Carl Sandburg.

ABOVE Advection fog over Hong Kong, spotted by David Law from the Pok Fu Lam Reservoir Trail.

OPPOSITE Saussure's cyanometer, designed to measure the blueness of the sky, now lives inside the Bibliothèque de Genève, Switzerland.

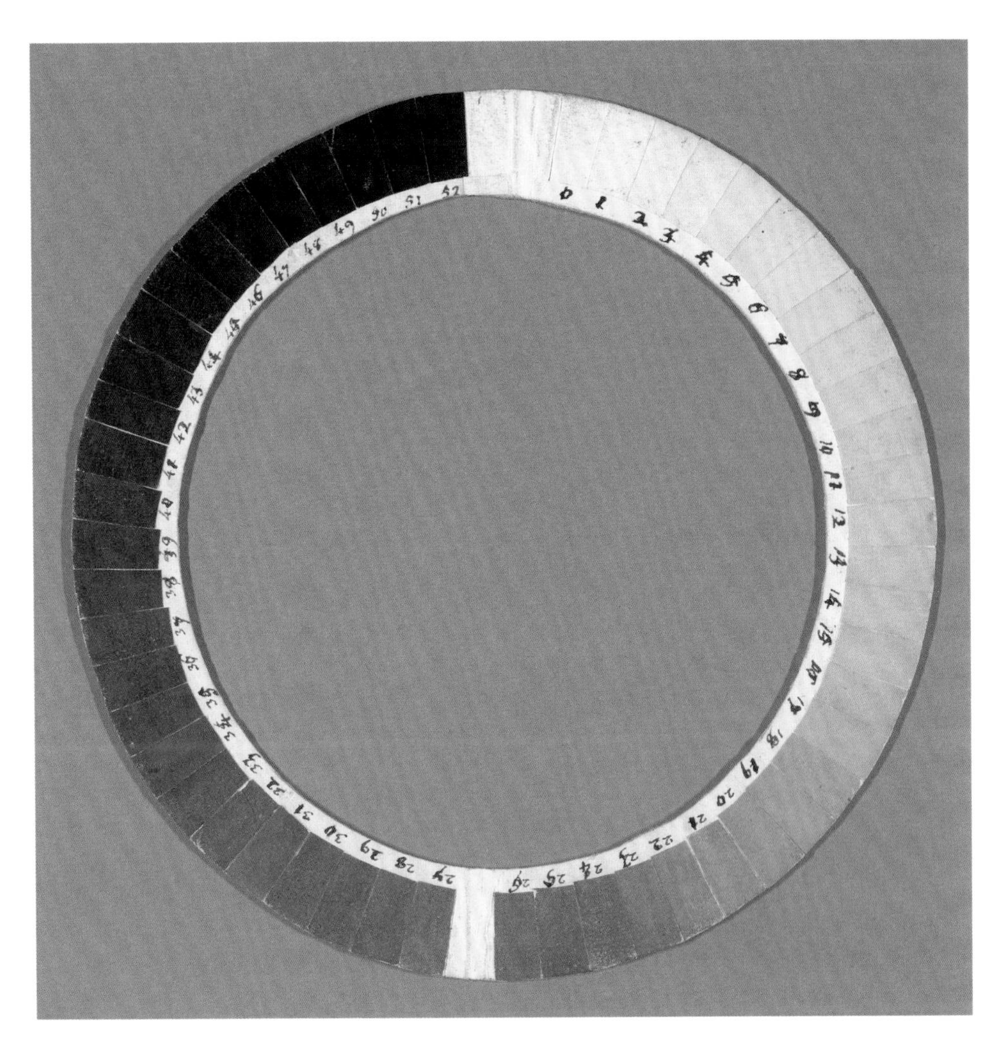

THE CYANOMETER was invented in 1789 by Horace-Bénédict de Saussure, a Swiss physicist and mountain climber, to aid his investigations into the blueness of the sky. Saussure dyed paper squares with different mixtures of Prussian blue and black inks in an effort to create every distinguishable shade of blueness, from white through to black. He assembled the 53 numbered cards into a circle that could be held up at a standard distance from the eye. The shades were used to give a number to the cloudless sky. Around Geneva, the sky was pale compared with that at the summit of Mont Blanc, which he had climbed in 1787, when his cyanometer was a prototype of loose pieces, to find the sky was 39 degrees blue. De Saussure theorized that the color depended on water droplets and ice crystals in the atmosphere above. Turns out he was completely right.

ON THE MEANING of the impressionistic swirls that dominate the sky of Vincent van Gogh's masterpiece *The Starry Night*, some say they were inspired by early illustrations of galaxies observed through the new powerful telescopes of the era. Others speculate that the swirls represent the fierce mistral winds that blow relentlessly for much of the year over Saint-Rémy-de Provence, France, and which are thought to have triggered the artist's breakdown in the town's psychiatric asylum. We'd like to add a third possibility. Perhaps the artist's illness brought him a moment of cloudspotting clarity? Perhaps he was the first to depict the wild swirls of a fluctus, or Kelvin-Helmholtz, cloud? The formation, photographed here over New Zealand by Greg Dowson, results from horizontal vortices induced by shearing winds. The conditions can be perfect over hills and low mountains like Provence's Alpilles range, visible from Van Gogh's window at the Saint-Paul asylum.

OPPOSITE TOP Spotted over Christchurch, New Zealand, in February 2016 by Greg Dowson (Member 15,705). **OPPOSITE BELOW** Spotted over Saint-Rémy-de-Provence, France, in June 1889 by Vincent van Gogh (not a Member).

ABOVE Cumulus fractus takes a stand over Ontario, Canada, spotted by Althea Pearson (Member 38,865).

IT IS NOT OFTEN that a cloud is given a name, but this Cumulonimbus storm cloud has one. It is called Hector the Convector, and it forms practically every afternoon from September through to March over the Tiwi Islands, north of Darwin, Australia. The cloud is enormous, often reaching to altitudes as high as 20 kilometers (12 miles). Hector's name was given to it by Second World War pilots, for whom the steadfast dependability of its position made it a natural navigation marker that was visible from great distances.

ABOVE "Hector the Convector," spotted by Christopher Watson (Member 7,692) off the coast of Darwin, Northern Territory, Australia.

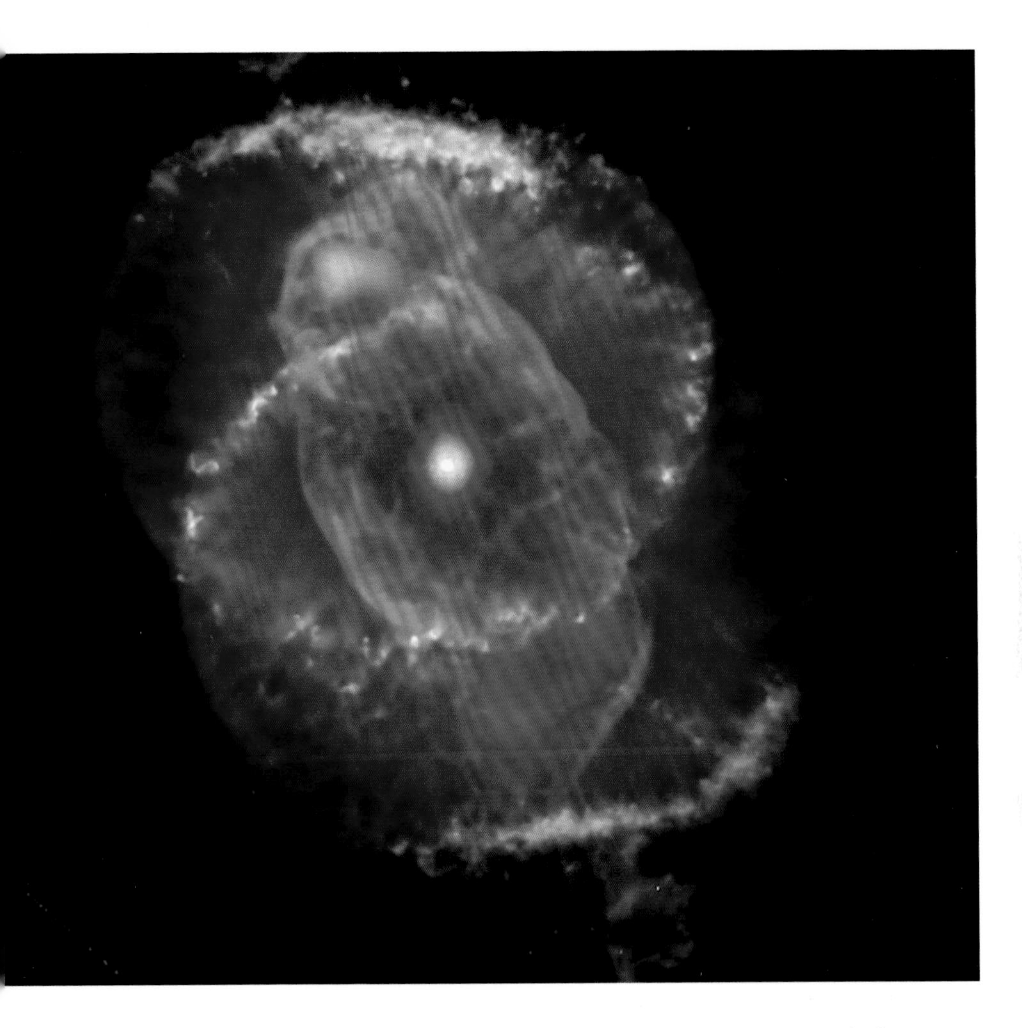

A NEBULA IS AN INTERSTELLAR CLOUD of dust, hydrogen, helium, and other ionized gases. This particular one, spotted by the Hubble Space Telescope, shows the effects of a hot, dying star and is nicknamed the Cat's Eye Nebula. It is visible from the northern hemisphere in the Draco constellation, between the Big and Little Dippers. The image was created by stitching together more than one image, using filters to recreate the colors as they'd appear to us from nearby.

ABOVE The Cat's Eye Nebula, spotted by the Hubble Space Telescope.

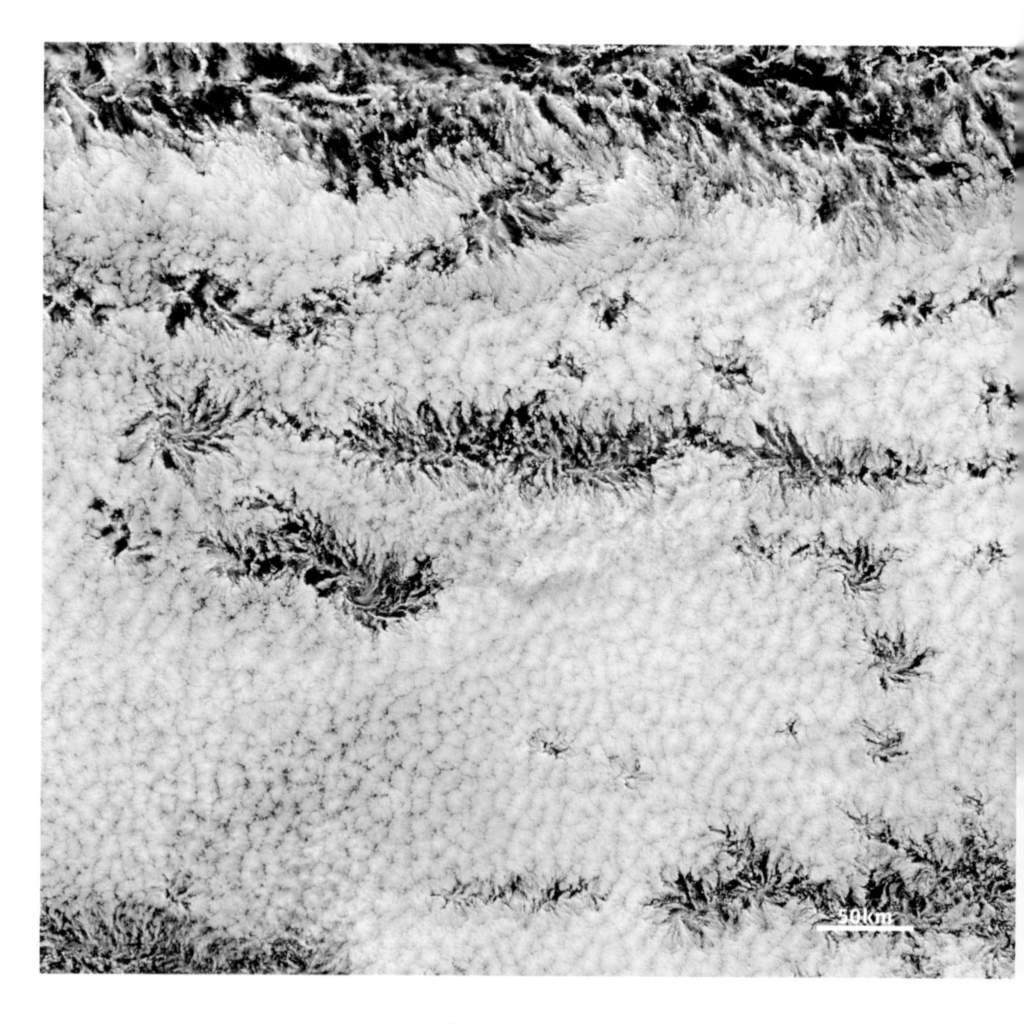

YOU'LL ONLY BE ABLE TO SPOT actinoform clouds if you're an astronaut—or a satellite. The radial, leaf-like patterns can form in Stratus and Stratocumulus layers over open oceans, but they are on such a huge scale that they can't be distinguished from below. The widths of the features are typically around 80–300 kilometers (50–180 miles). Actinoform clouds, whose name comes from the Greek word for "ray," were only identified in the 1960s when their distinctive patterns were spotted in early satellite photographs. We don't have a good understanding of why marine clouds should become arranged into these radial structures, nor why the patterns are in fact quite common.

Thousands of tired, nerve-shaken, over-civilized people are beginning to find out that going to the mountains is going home; that wildness is a necessity.

From *Our National Parks* (1901), by John Muir.

OPPOSITE Actinoform clouds, spotted within marine Stratocumulus by NASA's Terra satellite.

ABOVE Cirrus intortus clouds, spotted over Cape Town, South Africa, by Hetta Gouse.

THIS HAND-TINTED PHOTOGRAPH of Altocumulus clouds appeared in the first edition of the *International Cloud Atlas, a pioneering reference work for sky observers published in 1896, named the International Year of the Cloud.* It was put together by a team of early meteorologists who called themselves the Cloud Committee. Translated into English, French, and German, it included the main cloud types illustrated with loose-leaf prints held together in a ribbon-tied portfolio. The Atlas was the first example of photography being used as a reference for weather observers. Today, the *International Cloud Atlas* is published by the World Meteorological Organization. Its most recent 2017 edition was the first to appear online, some 121 years after the International Year of the Cloud.

ABOVE Altocumulus clouds, from the first edition of the *International Cloud Atlas.*

OPPOSITE A circumhorizon arc produced by a subtle layer of Cirrostratus cloud, spotted over Peacham, Vermont, US, by Mike Brown (Member 23,955).

WHEN YOU SEE WHAT LOOKS LIKE a rainbow that has been flattened into a horizontal line, you have likely spotted a circumhorizon arc. The optical effect is caused by ice crystals in the sky acting as tiny prisms that refract and reflect the light, and it appears as a bright, flat band of colors parallel to the horizon. Ranging from red at the top to indigo at the base, it appears well below a Sun that is high in the sky. In fact, a circumhorizon arc can only form when the Sun is very high—at an angle of more than 58 degrees above the horizon. This means that it is a summer optical effect, and one that can only form in some parts of the world. Even in midsummer, the Sun only rises high enough in the sky for a circumhorizon arc to be visible in latitudes below 55 degrees. As a result, this flat rainbow is common in Nairobi, infrequent in Seattle, and unheard of in Copenhagen.

THESE CLOUDS, DESCRIBED AT THE TIME as "Cumulo-stratus," were painted by Luke Howard, a pharmacist and Quaker who, at the dawn of the 19th century, became the man who named the clouds. Howard lived in London, where in 1802 he gave a talk to his scientific debating club in which he argued for a Latin system to classify clouds like the systems used for plants and animals. He proposed the names Cumulus, Stratus, and Cirrus as well as compound terms like Cirro-cumulus to refer to intermediate and transitional forms. Howard's talk and his articles and essays that followed founded the naming system for clouds that we still use today. He didn't paint this watercolor alone. The landscape was added by professional artist Edward Kennion, for this amateur meteorologist who forged a new language for the sky was rubbish at painting the ground.

ABOVE Rural landscape by Edward Kennion with cloud studies by Luke Howard (c. 1808–11).
OPPOSITE A pierced heart, spotted by Sim Richardson over Stansted Airport, Essex, England.

ANVIL CLOUDS, KNOWN OFFICIALLY AS INCUS, form in the upper parts of Cumulonimbus thunderstorms and spread out like, well, anvils as the storms' updrafts encounter the tropopause. This is the boundary region between the troposphere below and the stratosphere above, where the profile of the temperature change with altitude tends to stop rising air in its tracks. If the updrafts are powerful enough, however, they can punch through the invisible ceiling of the tropopause to form overshooting tops like these, spotted in 2016 by astronaut Tim Peake. The storms beneath overshooting tops tend to be severe. Updrafts with this much energy can support the weight of some serious hailstones.

ABOVE Incus or anvil clouds, spotted by astronaut Tim Peake aboard the International Space Station.

THE HORSESHOE VORTEX CLOUD is enigmatic and fleeting. It starts as a subtle, flat roll of cloud that forms within a horizontal vortex of air. The vortex occurs as an invisible rising column of air, known as a thermal, encounters crosswinds above and curls over in a spin. When conditions are just right, cloud droplets develop as temperatures drop within the low pressure along the center of the vortex. The thermal continues to push upwards from below, lifting the middle of the twisting roll of cloud, distorting it into a curve that resembles an upside-down horseshoe. Or maybe a floating mustache. Or Dracula's dentures.

ABOVE Horseshoe vortex cloud, spotted by Katrina Whelen over Bendigo, Victoria, Australia.

THE AMERICAN REALIST PAINTER Edward Hopper was a lover of high cloud. In his 1939 painting *Ground Swell*, he adorned the sky with jet-stream Cirrus, which reveal the ribbons of strong wind that encircle the globe in meandering paths, helping generate storms and steering them from west to east across temperate zones. Intense wind shear along the edges of the jet stream can corral these high clouds into stripes or patches, arranged in rows down the length of the jet stream. "My aim in painting," wrote Hopper in 1933, "has always been the most exact transcription possible of my most intimate impression of nature."

ABOVE Detail of *Ground Swell* (1939), by Edward Hopper.

ALL RAINBOWS HAVE THE POTENTIAL to be complete circles. We only see them as semi-circular bows because the Earth obscures the lower half. You need to be viewing the bow from above to see the whole thing—from a tall building, say, or a cliff edge. Or even, as Debra Ceravolo found, from the window of a small aircraft flying over the forested wilderness of Canada.

ABOVE The part of the rainbow you never normally see, spotted by Debra Ceravolo over eastern Ontario, Canada.

THIS BUSY SKY would be described as Stratocumulus castellanus and Cirrostratus that might have developed at the top of a distant Cumulonimbus. Thrown in for good measure are also the optical effects known as crepuscular rays. The Romantic poet Percy Bysshe Shelley put it another way in his 1813 poem *Queen Mab* as "far clouds of feathery gold / Shaded with deepest purple, gleam / Like islands on a dark blue sea."

ABOVE A mixed sky spotted over Colombey-les-Choiseul, Haute Marne, France, by Karin Enser (Member 43,050).

ABOVE A lightning man caught tip-toeing across the Bahamas, spotted by Michael Sharp (Member 19,947). Left: Delicate bands of Cirrus sweep like brushstrokes on a canvas, spotted over the west coast of southern Africa by Commander Alexander Gerst aboard the International Space Station.

CHAOTIC, FRACTURED SKIES can appear when Nimbostratus, the thick, grey layer of rain cloud, begins to break up and allow the blue to peek through. Such is the sky depicted here by the French Post-Impressionist painter Henri Rousseau. The featureless dark grey Nimbostratus blanket has torn into shreds and tufts at different altitudes. The way Rousseau painted the landscape gives the impression of high-visibility conditions. This is not just Rousseau's painting style. The lower atmosphere appears far less hazy after the sort of prolonged rain that falls from Nimbostratus. Airborne dust is swept to the ground by all the precipitation falling from the rain clouds. This is when clouds behave like nature's air filters. After the rain has gone and the Nimbostratus has broken into Stratocumulus, autumn colors appear all the more vivid through the cool, crystal air.

ABOVE *View of Bievre-sur-Gentilly* (c. 1895) by Henri Rousseau.

ABOVE Capricious Cirrus clouds dance across the heart of Australia. Spotted by Chris Devonport over Alice Springs, Northern Territory, Australia.

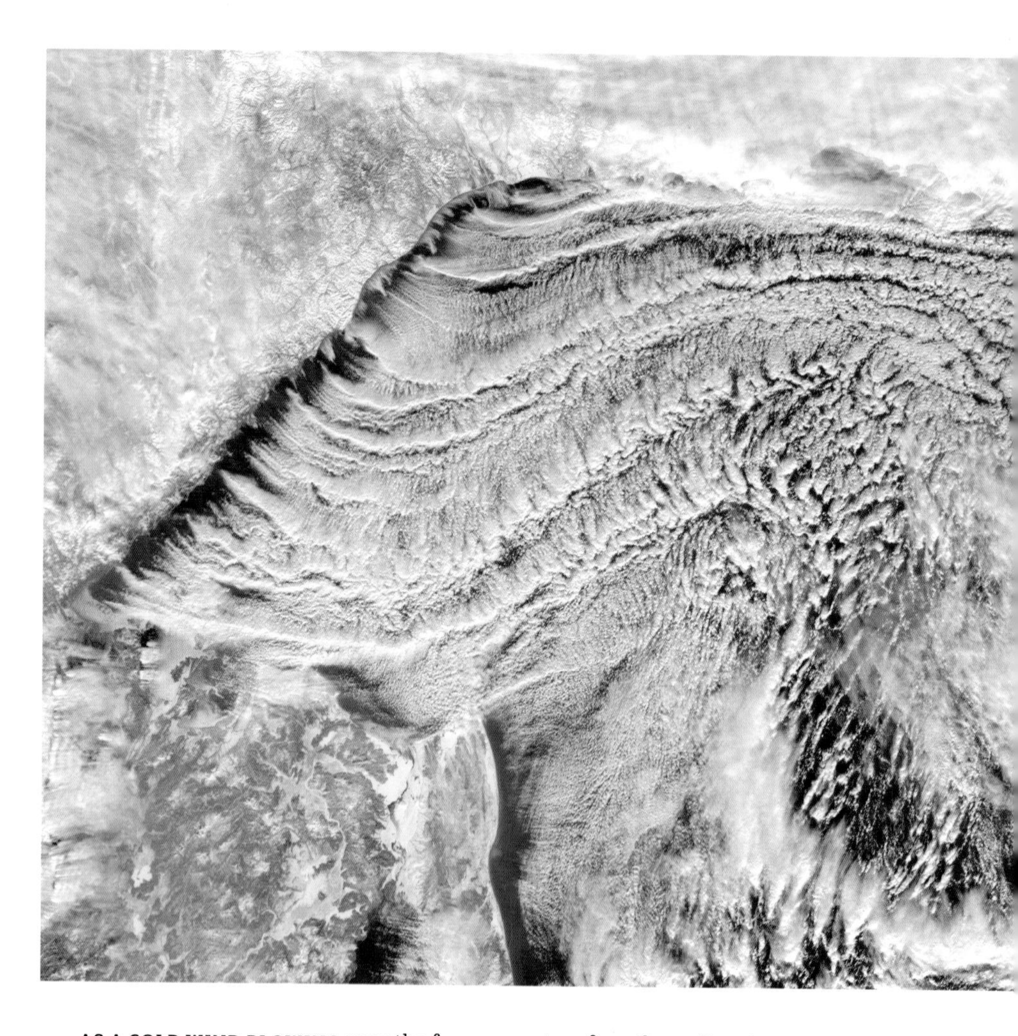

AS A COLD WIND BLOWING over the frozen wastes of northeast Russia encounters the Sea of Okhotsk, it forms sweeping lines of cloud known as cloud streets. The sea is distinctly warmer than the land, and so the low air in contact with it warms and picks up moisture. Unable to float upwards en masse through the colder air above, it arranges itself into cells of rising air where cloud forms, separated by regions of sinking air where the sky is clearer. The broad flow of the wind draws out this pattern into long orderly streaks of convection that are aligned along the direction of the wind. Viewed from below, cloud streets like this are known by the term radiatus, as the effect of perspective makes the parallel lines of cloud appear to radiate from a point on the horizon. From above, like this satellite view, cloud streets are the multilane highways of the sky.

ABOVE Cloud streets over the Sea of Okhotsk, Eastern Russia, spotted by NASA's Terra satellite.

They say that the Dead die not, but remain
Near to the rich heirs of their grief and mirth.
I think they ride the calm mid-heaven, as these,
In wise majestic melancholy train,
And watch the moon, and the still-raging seas,
And men, coming and going on the earth.

From "Clouds," *Collected Poems* (1916), by Rupert Brooke.

ABOVE Cumulus at sunset over Ocean Beach, San Francisco, US, spotted by Joan Laurino (Member 37,460).

THE CIRCUMZENITHAL ARC, sometimes referred to as "a smile in the sky," is an optical effect that looks rather like an upside-down rainbow. It is actually a type of halo phenomenon, which means it is created not by raindrops but by the refraction of sunlight through a cloud's ice crystals. The colors of the circumzenithal arc are brighter and more intense than those of a rainbow. You might wonder why more people don't know about this striking optical effect. One reason it is not better known is due to where it forms in the sky. The circumzenithal arc appears high overhead, rather than down near the horizon like a rainbow. Since it is positioned as the segment of a circle drawn around the zenith, this optical phenomenon is one that is generally missed—except, of course, by cloudspotters.

ABOVE Comfortable in its duvet of Stratus undulatus, Lolo Peak in Montana, US, has no plans to rise early this morning. Spotted by Michael Schwartz (Member 26,947).

OPPOSITE Circumzenithal arc formed by ice crystals in a subtle layer of Cirrostratus cloud, spotted over Los Angeles, California, US, by Christine Alico (Member 30,559).

ONE OF THE MORE VEXING CHALLENGES for cloud geeks is distinguishing between two different formations that both consist of parallel lines of cloud. These are fibratus, shown here on the left, and undulatus, on the right. Both are arranged into lines, so how do you know which is which? The essential difference is the orientation of the parallel cloud features compared to the wind direction. In fibratus, the filaments run parallel to the wind up at cloud level. For undulatus, the ridges run perpendicular to the wind. This difference has an effect on the cloud's appearance. For fibratus, think of strands of hair flowing in the wind. For undulatus, think of waves on the ocean—or perhaps those ridges of sand you feel beneath your bare feet as you take a stroll along the ocean's shore.

ABOVE LEFT Cirrus fibratus, spotted by Badr Alsayed over Tabūk, Minṭaqat Tabūk, Saudi Arabia.

ABOVE RIGHT Altocumulus undulatus, spotted over Puurs, Belgium, by Alain Buysse.

THE CHURCH OF ST. PETER AND ST. PAUL in Godalming, Surrey, England, was built around the year 1100, just after the Norman conquest, so it has survived for over 900 years. Floating above it, as if perched on the spire, is a Cumulus cloud. It is the species known as fractus because its edges are fraying as it dissipates into the blue. The cloud will have survived to the ripe old age of about ten minutes.

ABOVE Cumulus fractus, spotted over Godalming, Surrey, England, by Matthew Curley.

THE WINDOW OF AN AIRCRAFT always offers a new perspective on clouds. The American painter Georgia O'Keeffe was particularly inspired by them as a plane passenger in the late 1950s. She painted a series called *Above the Clouds* based on the views from aircraft windows. Although this first one, called *Above the Clouds I*, was on a standard-sized canvas (0.9 × 1.2 meters / 3 × 4 feet), O'Keeffe's desire to capture the huge expanse of the clouds viewed from above like this led her to paint *Above the Clouds IV* on a canvas 7.3 meters (24 feet) wide. That painting was meant to be exhibited at the San Francisco Museum of Art (now the San Francisco Museum of Modern Art) in 1970, but it couldn't be shown there. It wouldn't fit through the museum doors.

ABOVE *Above the Clouds I* (1962/1963), by Georgia O'Keeffe.

THE HANGING TENDRILS from these jellyfish clouds are known as virga. They are streaks of falling ice crystals that evaporate away in the warmer, drier air below. Were such showers reaching all the way to the ground, they'd be known as praecipitatio. Virga, by contrast, are when the tendrils float free, drifting in the calm currents of this ocean of air that we call our atmosphere.

ABOVE Virga trailing from Altocumulus clouds, spotted over Atrani, Italy, by Frieder Wolfart (Member 42,997).

Once I beheld a sun,
a sun which gilt

That sable cloud, and
turned it all to gold.

From "Night VII" in *Night Thoughts* (1742–1745), by Edward Young.

ABOVE Crepuscular rays at sunrise spotted over the Oxley Range, near Bourke, New South Wales, Australia, by Frank Povah (Member 46,285).

A HAZY FIELD OF CIRROSTRATUS with isolated patches of Altocumulus spread out over the undulating sands of Erg Chebbi in the Sahara desert. It is reassuring to know that even in the most oppressively hot summer climates, cool serenity can still be found in the sky.

ABOVE Cirrostratus and Altocumulus, spotted over the dunes of Merzouga, Morocco, by Jelte Vredenbregt (Member 35,462).

THIS GLOWING SPECTER could easily be misconstrued as a supernatural entity, but its true origin is simple. It is a sub-sun, a reflection of the Sun from the ice-crystal fog known as diamond dust. The twinkling, low-level crystals form in calm, extremely cold conditions—temperatures of perhaps -20°C (-4°F) or below—when the water vapor in the air freezes into a subtle, tumbling confetti of ice. Diamond dust shares a lot in common with Cirrus clouds—a lot, that is, except altitude—and like its high Cirrus cousins, this low ice fog can produce a range of different optical effects as its crystals bend and reflect the sunlight. A sub-sun can appear when you look down onto diamond dust. It is a blurred image of the Sun, formed by the collective sparkles glinting from the tops of countless tiny mirrors of ice.

THIS CLOUD IS CIRROCUMULUS. It is the least common of the ten main cloud types. Like the low-level Stratocumulus and mid-level Altocumulus layers, Cirrocumulus is composed of many individual cloud clumps, known as cloudlets. These are in fact the same size as the larger-looking cloudlets of the lower clouds. Cirrocumulus are just much higher in the sky, at altitudes perhaps of 8 or 9 kilometers (5 or 6 miles), and so from all the way down here, its cloudlets appear no larger than tiny celestial grains of salt.

ABOVE Cirrocumulus, spotted over Northumberland, England, by Maria Wheatley (Member 23,424).

OPPOSITE A sub-sun spotted in diamond dust over Walnut Canyon, Arizona, US, by Tom Bean (Member 41,135).

ABOVE Crepuscular rays formed as the office lights from the windows of the Bank of America Center building are scattered by the city's finest low-lying Stratus, aka fog. Spotted by Matt Friedman, San Francisco, California, US.

OPPOSITE *A Balloon Prospect from Above the Clouds* was included in Thomas Baldwin's book *Airopaidia* (1786).

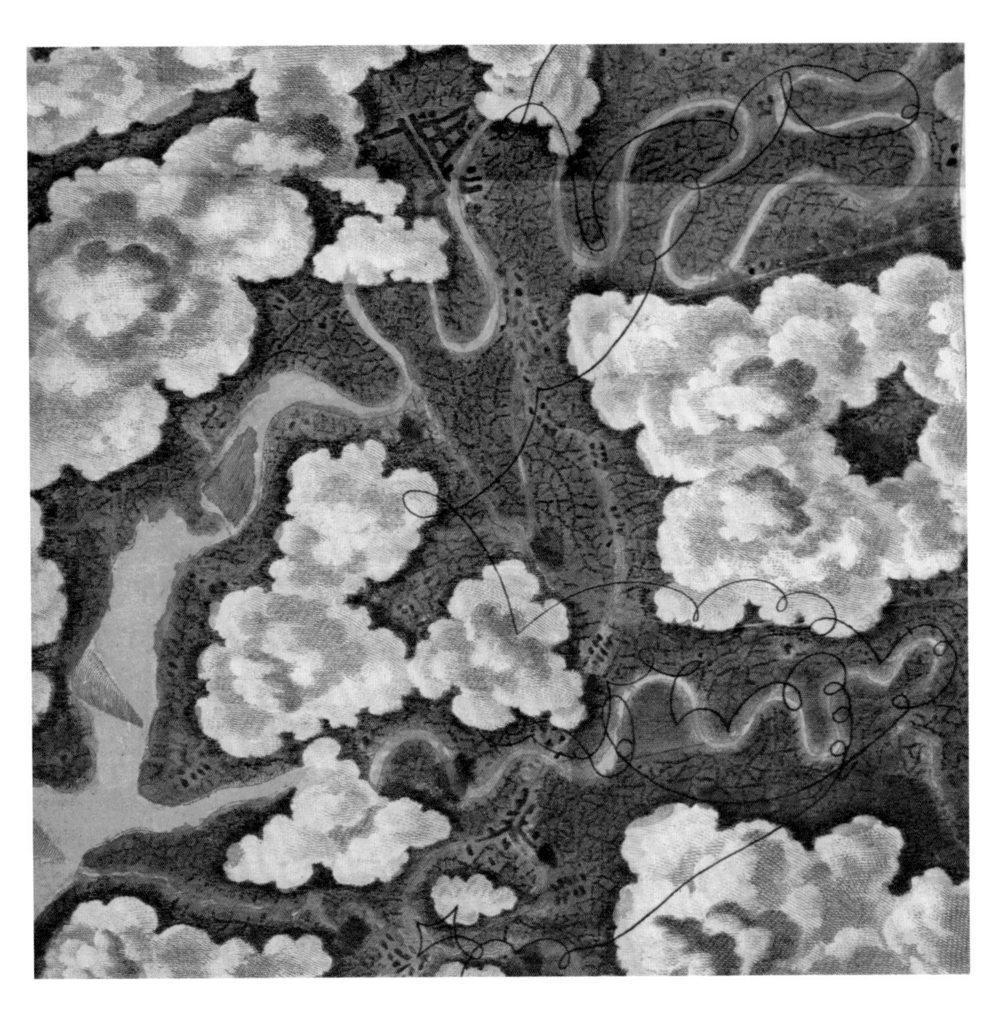

THIS IS BELIEVED TO BE THE EARLIEST printed representation of the view looking down from the basket of a hot-air balloon. The scene was observed during a 1785 flight over Cheshire and Lancashire, England, by a clergyman called Thomas Baldwin. He had long dreamed of taking to the sky in a balloon, and his chance came when the Italian balloonist Vincenzo Lunardi came to town and was unable to make the planned balloon flight due to an injury. Baldwin offered to stand in and later published a detailed account of the experience along with engravings based on his sketches. Looking down on the fair-weather Cumulus clouds, with rivers and lakes in pink below, the revolutionary bird's-eye view has superimposed over it a looping, swirling black line. This represents the route of the balloon as it was swept along by the capricious motions of the wind.

In the immense void of the sky, clouds arise.

They come from nowhere and they go nowhere.

Nowhere exists in a storehouse of clouds.

They arise in the empty spaces of heaven and dissolve, therein, like thoughts in the human mind.

The Buddha, quoted in *The Wisdom of Water* (2008), by John Archer.

ABOVE The Buddha, also known as a towering Cumulus cloud, spotted over Naples, Florida, US, by Julie Magyar Africk (Member 41,630).

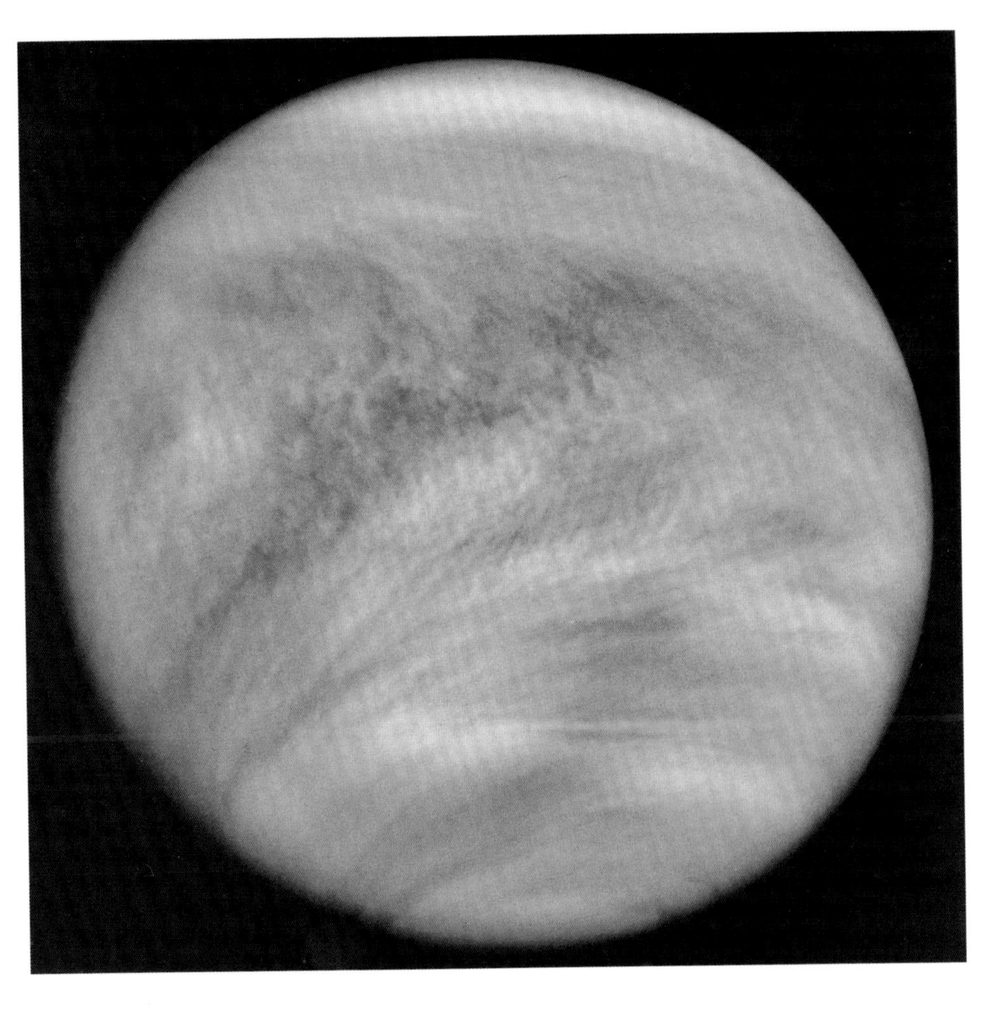

NEXT TIME SOMEONE COMPLAINS about a grey, overcast Altostratus sky, tell them to be thankful we don't have Venus's clouds. Our neighboring planet is wreathed in near-permanent clouds of pure sulphuric acid. These are suspended in a carbon dioxide atmosphere that is 90 times denser than Earth's. Just 10 percent of sunlight hitting these clouds ever reaches through to the surface, but all that CO2 has a huge greenhouse effect. It traps in the Sun's heat so effectively that temperatures on Venus's surface exceed 450°C (842°F). The only place in our solar system hotter than Venus is on the surface of the Sun itself.

ABOVE Cloud structures in the Venusian atmosphere, revealed in ultraviolet wavelengths of light and spotted in 1979 by NASA's Pioneer Venus Orbiter.

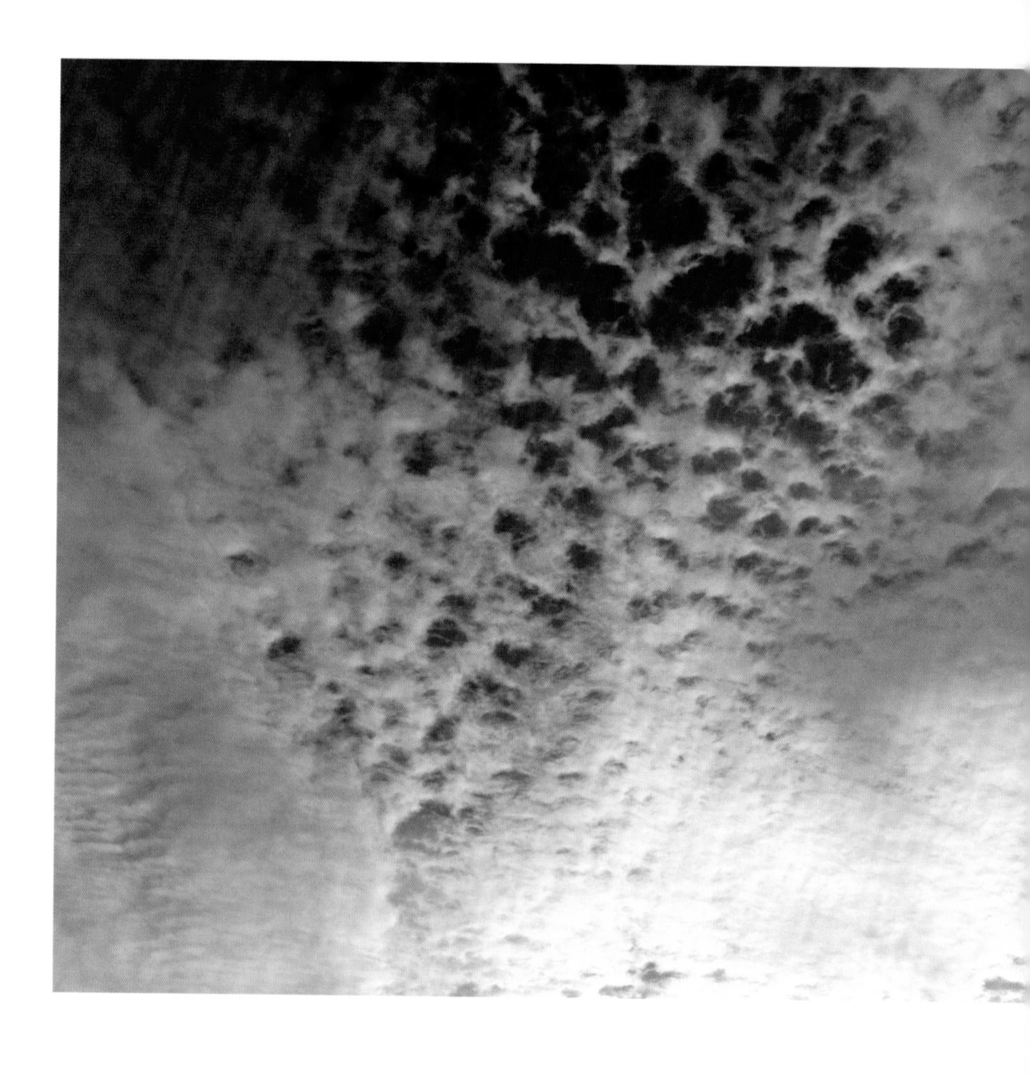

ABOVE Lacunosus clouds, derived from the Latin for "full of holes"—like gazing up at soap suds beneath a warm bubble bath. Spotted over Bloemendaal, Netherlands, by Hans Stocker (Member 36,089).

OPPOSITE An Altocumulus lenticularis cloud, spotted by Terence Pang as it takes a dive into Hong Kong's Victoria Harbour.

ICE CRYSTALS ABOVE. Ice crystals below. Some in moonlit Cirrus. Some in fresh-laid snow.

ABOVE An Altocumulus lenticularis cloud, spotted by Terence Pang as it takes a dive into Hong Kong's Victoria Harbour. Detail from *Winter Scene in Moonlight* (1869), by Henry Farrer.

OPPOSITE LEFT Altocumulus lenticularis over the Sawatch range, Colorado, US, spotted by James Brooks (Member 44,546).

OPPOSITE RIGHT Solomon R. Guggenheim Museum, New York, US, spotted by Jean-Christophe Benoist.

THE AMERICAN ARCHITECT Frank Lloyd Wright hated cities, and when the mining magnate Solomon R. Guggenheim prevailed upon him to design an art museum in Manhattan, Wright seems to have taken inspiration from the clouds. Stacked forms of Altocumulus lenticularis are a fairly common sight in Arizona, where Wright had established his school of architecture. Is it a coincidence that the profile of the Guggenheim bears such a resemblance to these clouds? We think not. Wright and Guggenheim argued over the color for the museum's exterior. The architect eventually backed down, but his preferred color had been a Cherokee red. There's no record of clouds being referenced in any of the discussions, but we can't help noticing how, as dusk approaches, lenticularis clouds take on the warm hues of the setting Sun—oranges, pinks and, yes, Cherokee red.

ABOVE An eye in the sky, spotted over Torrejón de Ardoz, near Madrid, Spain, by Adolfo Garcia Marin (Member 12,115).

Those who dream by day are cognizant of many things which escape those who dream by night.

From Eleanora (1842), by Edgar Allan Poe.

ABOVE A quilted blanket of Altocumulus at sunset over Penge, London, England, spotted by Christina Brookes (Member 33,764).

THE AURORA BOREALIS AND AUSTRALIS appear in the higher latitudes when charged particles from the Sun interact with those trapped within the Earth's dynamic and shifting magnetic field. As the particles are drawn into the upper atmosphere by the magnetic field, they excite oxygen and nitrogen atoms, which they cause to emit light. The stream of charged particles from the Sun is called the solar wind, and it stretches out the Earth's magnetic field like an invisible windsock. Aurora activity spikes after the wind from large bursts of solar emission, known as coronal mass ejections, stretches the field so far into space that it eventually snaps back, pulling a flood of charged particles in with it. This display was spotted by astronaut Alexander Gerst. He remarked that even though the lights are visible regularly from space, they are still "mind-blowing, every single time."

ABOVE A Cumulonimbus shower cloud shows us how it's done. Spotted by Sylvia Fella (Member 42,586), Eastbourne, East Sussex, England.

OPPOSITE The dancing lights of the aurora, spotted by astronaut Alexander Gerst while aboard the International Space Station.

SHROUDED IN A HUGE CAP cloud with a lenticularis above and a layer of Stratocumulus below, the Mayon Volcano in the Philippines is clearly not up for facing the tourists today.

ABOVE A volcano shroud cloud, spotted by Chito L. Aguilar from his balcony in Daraga, Albay, the Philippines.

OPPOSITE Throwing a ball through a hoop over Paterson, New Jersey, US, spotted by Edward Hannen. Also known as a horseshoe vortex cloud and Cumulus humilis.

CLOUDS CAN BE MADE FROM many things other than water. In this satellite image huge swirling clouds consisting of billions of tiny phytoplankton trace the currents and eddies of the Black Sea. But when they form these blooms, the microscopic aquatic organisms don't just mirror atmospheric clouds, they can also help to create them. Scientific studies in the Southern Ocean have found that gases and particles released into the air by marine phytoplankton can significantly contribute to cloud cover by acting as the "seeds" onto which the water droplets of clouds start to condense.

ABOVE The Black Sea with algal blooms, spotted by NASA's Aqua satellite in 2017.

OPPOSITE Cumulus reflections, spotted on the shore at Oostduinkerke-Bad, Flanders, Belgium, by Tim De Wolf (Member 38,959).

I never saw a man who looked
So wistfully at the day.
I never saw a man who looked
With such a wistful eye

Upon that little tent of blue
Which prisoners call the sky,
And at every drifting cloud that went
With sails of silver by.

From *The Ballad of Reading Gaol* (1898), by Oscar Wilde.

EVER WONDERED what a sunset looks like on Mars? Of course you have! Here is one viewed from the planet's Gusev Crater. In a reversal of our own sunsets, the Martian sky glows a pale blue around the setting Sun, while elsewhere it has its usual daytime color of butterscotch. Twilight also lasts longer on Mars. All the dust in the Martian atmosphere continues to scatter the light for as long as two hours after the Sun has dipped away from view.

ABOVE NASA's Spirit exploration rover enjoys the sunset on Mars.

ABOVE A Cumulonimbus, spotted by Andy Sallee (Member 37,600), seems to have taken particular dislike to this borough of Orlando, Florida, US.

WITH ITS DULL, GREY, FEATURELESS APPEARANCE, the mid-level layer cloud known as Altostratus is generally considered to be the boring cloud. But when the light is right, every cloud has its moment to shine.

ABOVE Altostratus, spotted at sunset over Mount Pleasant, Falkland Islands, by Melyssa Wright (Member 23,652).

THESE COLORED RINGS can appear as your shadow is cast onto a layer of cloud or fog. The optical effect is known as a glory, and is caused by the diffraction of sunlight as it reflects back off the cloud's tiny droplets. The phenomenon is also called a Brocken spectre, after the Brocken mountain in Germany, where it is often seen by walkers climbing through low cloud into the sunlight above. The shadow's blurred edges and its distortion due to the effects of perspective account for the Brocken spectre's ghostly appearance.

ABOVE A glory, or Brocken spectre, spotted by Tom Bean (Member 41,135) from the rim of Anderson Mesa, south of Flagstaff, Arizona, US.

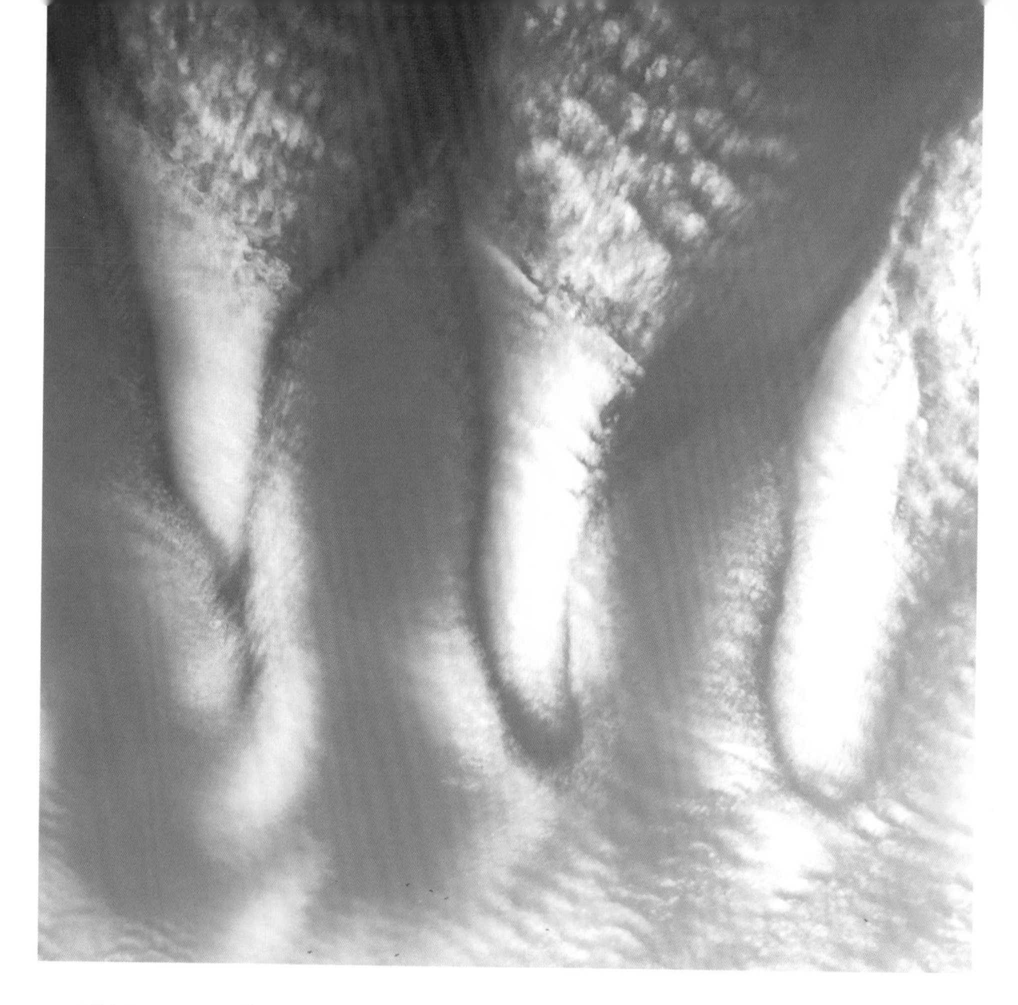

Sit in reverie and watch the changing color of the waves that break upon the idle seashore of the mind.

From *The Spanish Student: A Play, in Three Acts* (1843), by Henry Wadsworth Longfellow.

ABOVE Undulations within undulations as mountain waves shape Cirrocumulus clouds, spotted over Lossiemouth, Moray, Scotland, by Melyssa Wright (Member 23,652).

OPPOSITE LEFT Cumulonimbus calvus form, spotted over Madrid, Spain, by Antonio Martin (Member 11,271). OPPOSITE RIGHT Cumulonimbus capillatus form, spotted over Bukovník, Czech Republic, by Karel Jezek (Member 34,987).

WHEN YOU ARE ABLE TO OBSERVE a Cumulonimbus storm cloud from afar, you can often see that it has a distinctive shape, spreading outwards at the top to resemble a mushroom or a blacksmith's anvil. The canopy of cloud above, known as incus, is the part to pay attention to when identifying a Cumulonimbus. A Cumulus cloud has developed into a Cumulonimbus when its top has glaciated. This means that the droplets in its upper reaches have begun to freeze, making the edge of the cloud's summit appear blurry. The cloud above left is described as Cumulonimbus calvus, from the Latin for "bald," because the top has begun to glaciate, its edge softening but still fairly smooth. The more mature one above right is described as Cumulonimbus capillatus, meaning "hairy," as its whole upper region has frozen into ice crystals, making the top much fluffier.

ANY CLOUDSPOTTER WHO VISITS the Storkyrkan cathedral in Stockholm, Sweden, will be lucky enough to see a whole range of the many arcs, rings, and spots of light that can appear when sunlight passes through the pure, prismatic ice crystals of clouds. These optical effects appear in a painting, dated 1636, by Jacob Elbfas, called *Vädersolstavlan*, which is the Swedish for "The Sun Dog Painting." Sun dogs, the spots of light that can appear on one or other side of the Sun, are just one of many halo phenomena formed as sunlight is reflected and refracted through a cloud's hexagonal ice crystals. The painting depicts a dramatic halo display over Stockholm on April 20, 1535, although not all the effects could have appeared at the same time. The rings, arcs, and lines of light play the starring role in the world's earliest surviving painting of halo phenomena.

CLOUDS HELP US UNDERSTAND the complex movements of our atmosphere by revealing the otherwise invisible wind patterns and temperature changes. Sometimes the signs are subtle, such as gentle ripples dancing across high cloud layers. At other times, though, they can be bluntly obvious, like the way these Altocumulus clouds show with crisp precision the back edge of a departing weather front.

ABOVE A divided sky, spotted over Ipswich, Suffolk, England, by Kenneth R. Carden (Member 20,402).

OPPOSITE *Vädersolstavlan* (1636), by Jacob Elbfas after the original (c. 1535) by Urban Målare.

A thrill of thunder in my hair:
Though blackening clouds be plain,
Still I am stung and startled
By the first drop of the rain:

From "A Second Childhood" (1922) in *The Ballad of St. Barbara and Other Verses*, by G. K. Chesterton.

ABOVE One of those ribbon-swirling gymnasts, spotted mid-routine by Jente De Schepper over Leuven, Flanders, Belgium.

OPPOSITE Cumulonimbus clouds bring a moist morning to Manhattan, New York, US. Spotted by Maxine Hill (Member 43,765).

IN THE HEAT OF THE DAY, a scattering of Cumulus clouds can sometimes mirror the shape of coastlines and islands. Like these over the Galápagos Islands off the coast of Ecuador, Cumulus form on columns of air, or thermals, which rise from the ground as they heat up in the Sun. Temperatures on the ocean remain steadier than on the land, and so offshore thermal activity and its corresponding cloud formation is suppressed. No doubt, all this will have been going through the mind of the cloudspotting astronaut when he took this photograph out of the window of the International Space Station.

ABOVE Cumulus clouds over the Galápagos Islands, spotted by the crew of the International Space Station.

OPPOSITE The Rainbow Serpent depicted in rock art in the Kakadu National Park, Northern Territory, Australia.

FOR EARLY ABORIGINAL CULTURES across Australia, the Rainbow Serpent represented rain and fertility. Depicted here on Ubirr Rock in the Kakadu National Park of the Northern Territory, the Serpent was one of the creation-beings. She was considered responsible both for providing life-giving water and also for bringing storms and floods to punish those who upset her. When a rainbow appeared in the sky, the Rainbow Serpent was moving from one waterhole to another. She was the reason some waterholes remained wet even in severe droughts. There is no consensus on the age of this painting, but the Serpent began to appear in the rock art of this region around 6,000 years ago.

LIKE ANY SENSIBLE LANDSCAPE PAINTER, the English artist J. M. W. Turner made numerous cloud sketches. The blue paper of this sketchbook, made towards the end of the 18th century, feels perfect for a scene of clouds illuminated by moonlight and reflected in the waters below. Turner was to develop into a master of capturing the moods of the sky. "Indistinctness is my forte," is how he explained it to a friend 50 years later.

ABOVE A cloud sketch in J. M. W. Turner's 1796 *Studies near Brighton Sketchbook*.

OPPOSITE Not all lenticularis clouds are shaped like UFOs or the lentils from which their name derives. This one, spotted by Patrick Dennis (Member 43,666) over Boulder County, Colorado, US, is an effervescent tablet dropped in a glass of water.

IT WAS A BIG DAY for the Cloud Appreciation Society when the World Meteorological Organization (WMO) finally accepted asperitas as a new official classification of cloud. Back in 2008, we proposed that this formation should be recognized as a cloud type in its own right. We had noticed its distinctive features, characterized by turbulent, chaotic waves that sometimes descend in peaks, in photographs sent by members around the world. To be considered official, a classification needs to be included in the *International Cloud Atlas*. This is the definitive reference work on the naming of clouds, which first appeared in 1896. The *Atlas* is now published by the WMO, which launched the online edition in 2017, including asperitas for the first time. The cloud's name comes from the Latin for "roughness," and it appears like a chaotic, turbulent sea seen from below. When it was accepted as official in 2017, asperitas became the first new classification of cloud in 54 years.

LEFT Asperitas, spotted over Erm, Netherlands, by Nienke Lantman (Member 24,009).

AIRCRAFT CONDENSATION TRAILS, or contrails, are the man-made clouds caused by the condensing, sometimes freezing, of the water vapor in aircraft exhaust. The shadow from a contrail can be made visible by a layer of cloud or haze in the atmosphere below subtly scattering the light. It might be a Cirrostratus cloud, barely noticeable as a milky whitening of the blue. But the contrail shadow spotted here by Daniel Fox over New Mexico, US, is more likely to be showing due to dust blown up from the dry ground. From where Daniel was walking his dog, Lucy, the trail and its shadow just happened to be perfectly aligned.

ABOVE A contrail casting its shadow onto atmospheric haze, spotted over Las Cruces, New Mexico, US, by Daniel Fox (Member 40,744).

Big whirls have little whirls, that feed on their velocity; and little whirls have lesser whirls, and so on to viscosity.

From *Weather Prediction by Numerical Process* (1922), by Lewis Fry Richardson.

ABOVE Fluctus, also known as Kelvin-Helmholtz wave clouds, spotted by Lana Cohen over Ny Alesund, Svalbard, Norway.

VAST INTERSTELLAR CLOUDS composed largely of hydrogen and cosmic dust are the spawning grounds for new stars. This one is shooting superheated gas out of its poles at 160,000 kilometers per hour (100,000 miles per hour). Anyone who complains when Earth's clouds get in the way of our Sun should remember that clouds made the Sun in the first place.

SATURN HAS MANY MOONS, 53 of which have been given names. One of these, Enceladus, is coated in a thick crust of ice. The moon is lacking in what we would normally think of as clouds, but it does exhibit huge cloud-like plumes of icy particles and water vapor. These spew far into space from Enceladus's southern pole. The cause of the plumes is unclear, but they might originate from an underground sea. Heating by the moon's interior could cause such high pressure in this hidden reservoir that water erupts through the icy surface to produce enormous, extra-terrestrial cloud jets.

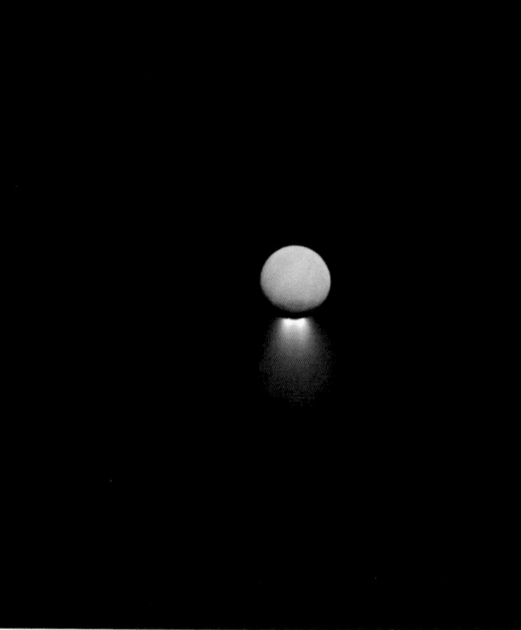

ABOVE Newborn star in a dust cloud, spotted by NASA's Goddard Space Flight Center using the Hubble Space Telescope.

RIGHT Jets of cloud emitted from the south pole of Enceladus, one of Saturn's moons, spotted by NASA's Cassini spacecraft.

OPPOSITE A parhelic circle and 120-degree parhelion caused by diamond dust, spotted by Richard Corrigall (Member 4,393) over La Plagne, France.

THE COLD WINTER AIR high in the French Alps is perfect for forming the glittering ice-crystal mist called diamond dust. When ice crystals grow slowly in drier conditions, they can form as tiny, clear, hexagonal prisms that bend and bounce sunlight to form all manner of curious spots and rings across the sky, known collectively as halo phenomena. The curved line of light appearing here is one such phenomenon, called a parhelic circle. It is a white band that runs parallel to the horizon, sometimes extending 360 degrees around the whole sky. It is always at the same height as the Sun, and tends to be a very fleeting effect. The bright spot appearing along the parhelic circle here is called a 120-degree parhelion. This is similar to a regular parhelion, known as a sun dog, but it is colorless, much further away from the Sun, and far less frequently spotted.

THE CLOUD SPECIES known as castellanus is not the most distinctive of formations. In fact, it is easily missed, even by cloudspotters. But castellanus clouds like these spotted by Sallie Tisdale over Oregon, US, are worth looking out for because they often forecast storms later in the day. The cloud's turrets indicate that the atmosphere up at the cloud level is unstable. When the crenellations of castellanus appear in Altocumulus clouds like these, they indicate that the unstable air is up at the mid-level of the clouds. This is significant. It suggests that any Cumulus clouds building from below as the day progresses will, upon reaching the unstable air, just keep growing. They'll likely continue to develop taller and taller until they've matured into Cumulonimbus storm clouds. "The sky was active all day," Sallie confirmed, "and that night we did indeed have thunderstorms."

WHEN THE CLOUDLETS in a layer of mid-level Altocumulus have gaps between them like this, they are described as perlucidus. This print by the 19th-century Japanese artist Katsushika Hokusai is from his classic series *Thirty-Six Views of Mount Fuji*. It is titled *South Wind, Clear Sky* (*Gaifū kaisei*). That strikes us as a bit of a misnomer.

ABOVE Altocumulus perlucidus, spotted over Mount Fuji, Japan, in around 1830 by Katsushika Hokusai.

OPPOSITE Altocumulus castellanus, spotted over northwest Oregon, US, by Sallie Tisdale (Member 42,126).

THESE TWO CLOUD CAPS, known as pileus, developed in Malawi over the top of Cumulus congestus clouds that were well on their way to maturing into Cumulonimbus storm clouds. Pileus can appear as airstreams overhead that are lifted by the Cumulus's rising convection currents, which can be extremely powerful in the tropics. The delicate-looking caps are like lenticularis or mountain cap clouds that have formed not over mountainous terrain but mountains of Cumulus. Droplets form where the airflow lifts and cools, only to evaporate away again as it sinks back down and warms beyond the cloudy obstacle. Being so short-lived, the droplets tend to be small and even in size—perfect for diffracting the sunlight and separating it into pearlescent colors. These can be just visible if you look carefully. Clearly, not every cloud has a silver lining. Some clouds have subtle multicolored ones.

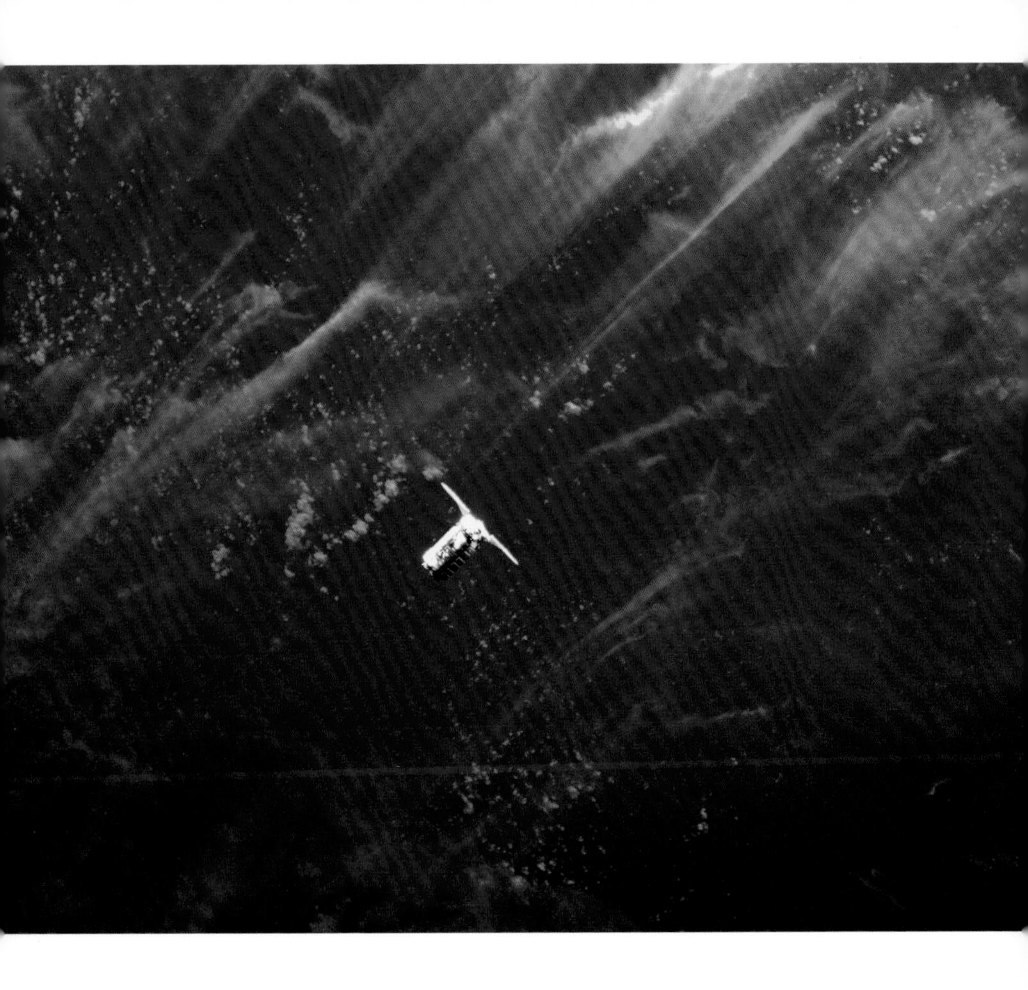

ABOVE A sunset of Cumulus, Cirrus, and Cirrus floccus clouds provides the backdrop as NASA's Cygnus spacecraft approaches the International Space Station for docking. The craft, photographed by ISS flight engineer Thomas Pesquet, is carrying supplies for the crew and equipment for their experiments.

OPPOSITE Pileus on top of Cumulus congestus, spotted by Jaap van den Biesen and Nienke Edelenbosch over Nkhata Bay, Malawi.

ABOVE *Cloud Gate* (2004), a public sculpture created by Indian-born British artist Sir Anish Kapoor in Millennium Park, Chicago, US, shines beneath Cirrus, a high cloud created by ice crystals cascading from the upper reaches of the troposphere.

A BUILDING WITH ITS HEAD IN THE CLOUDS is in the grey, featureless world of the Stratus cloud. Stratus is the lowest of the ten main cloud types. Though beautiful when viewed from above, it can feel close, even oppressive, from below. When it forms down at ground level Stratus is known as fog or mist, but up at 450 meters (1,500 feet) where it often lurks, it is the sky that skyscrapers scrape.

ABOVE Stratus, spotted by Filip Gavanski over the Roppongi district of Tokyo, Japan.

WE ARE USED TO SEEING CLOUDS turn pink at the beginning and the close of day as they reflect the rosy hues of the low Sun. But this Stratocumulus cloud over Utah, US, appeared pink at 1.00 pm. What on Earth could cause clouds to be colored like this at lunchtime? The cause is indeed found on Earth. The pink patch on the cloud's underside is a reflection of the Great Salt Lake below. Its waters can develop a distinctly reddish hue due to the salt-loving bacteria that thrive in its highly saline environment. They are tiny microbes with the power to summon sunsets in the middle of the day.

ABOVE A Stratocumulus sunset at lunchtime? Spotted in the middle of the day over Great Salt Lake, Utah, US, by Jeremy Hanks (Member 41,507).

ROLL CLOUDS, KNOWN OFFICIALLY AS VOLUTUS, form within a traveling wave of air. One can sometimes travel ahead of an advancing storm, but it is more commonly produced by the interaction of sea breezes. This is why roll clouds tend to be spotted over coastal waters. The rising air at the front of the wave and the dipping air at the back are not visible, just the rotating roll of cloud in the middle, but they can be felt as shifting air currents. "There was a sudden strong wind," reported Jean Louis Drye, who spotted the formation, "just as the cloud was passing over."

ABOVE A traveling volutus, or roll cloud, spotted by Jean Louis Drye as it sweeps over his cargo ship off the coast of São Paulo, Brazil.

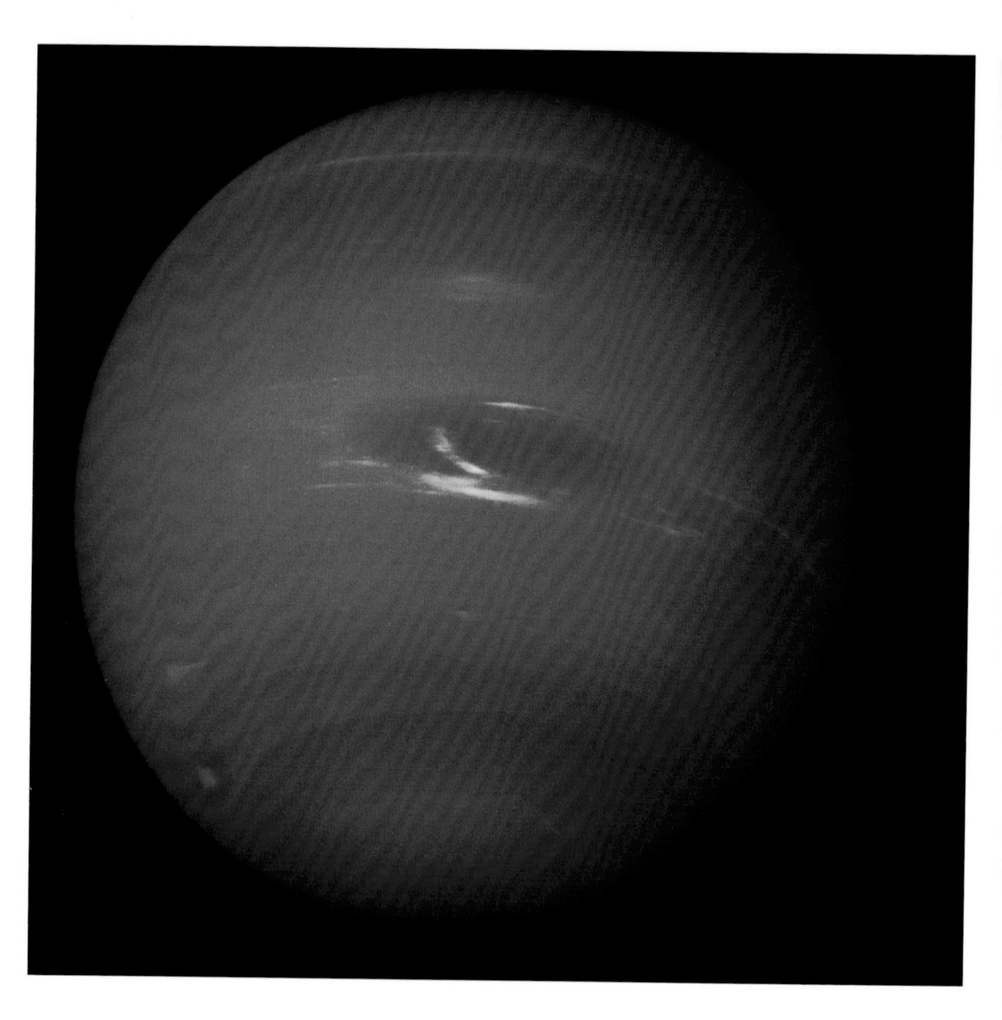

THE ICE GIANT NEPTUNE is the furthest known planet in our solar system, and has an atmosphere composed largely of hydrogen and helium. Neptune also has vast streaks of cloud that consist of frozen methane and ammonia. From space, Neptune's atmosphere appears quite serene, thanks to its bright blue color. It is, however, anything but calm. The Great Dark Spot in the center of this image is an enormous storm system extending the width of Earth that swirls around the planet, creating winds of 2,100 kilometers per hour (1,300 miles per hour). The isolated white patch appearing here a little below the Great Dark Spot is a region of cloud dubbed the "Scooter." This is because the persistent cloud feature rotates faster around the planet, lapping the Dark Spot every 12 rotations. The calm-looking Neptune turns out to have some of the strongest winds in the whole of our solar system.

I would build a cloudy House
For my thoughts to live in;
When for earth too fancy-loose
And too low for Heaven!

From "The House Of Clouds" (1841), by Elizabeth Barrett Browning.

ABOVE Altocumulus reflections, spotted by Maarten Hoek in Utengule, Mbeya, Tanzania.
OPPOSITE Neptune, spotted in 1989 from NASA's Voyager 2 spacecraft.

KNOWN AS THE "KING OF CLOUDS," the Cumulonimbus storm cloud can reach as high as 16 kilometers (10 miles), which makes it the tallest of all the cloud types. This is the formation that gives us the joyful phrase "on cloud nine." When the first edition of the cloud-identification manual the *International Cloud Atlas* was published in 1896, Cumulonimbus was ninth in the list of ten main classifications, known as the cloud genera. To be on cloud nine, therefore, is to be on the tallest cloud.

ABOVE Cumulonimbus, spotted over Darwin, Northern Territory, Australia, by Cecelia Cooke (Member 32,344).

OPPOSITE *Genesis, The Creation: Division of Sea and Earth* (c. 1467), spotted by Heather Silverwood (Member 42,036).

ANY GUESSES AS TO THE FAVORITE CLOUD formation of the medieval Dutch illuminator known as The Master of Evert Zoudenbalch? The leading illuminator in Utrecht towards the end of the 15th century, the artist's actual name is not known, but this name is used by some historians because he was the key artist for a richly illuminated Bible commissioned in 1460 by Evert Zoudenbalch. The skies throughout the artist's contributions to the Bible are filled with one particular type of cloud: Cirrus uncinus. This is a form of the high, ice-crystal Cirrus cloud that has a distinctive hooked appearance, often described as "mares' tails." The Master of Evert Zoudenbalch's love of Cirrus explains why other art historians choose to refer to him by a far better name: "The Master of Feathery Clouds."

LIKE A RIBBON OF LIGHT that has been tied in a bow, these rare halo phenomena are centered on the anti-solar point. This is the spot diametrically opposite the position of the Sun, which is shining here from behind the camera. The arcs of light are formed by the complex refraction and reflection of sunlight through tiny ice crystals in high cloud. The optical phenomena visible here are a subparhelic circle arching horizontally across the image, a diffuse arc appearing as the top loop of the bow, and a parry anti-solar arc appearing as the tails of the bow, with the anti-solar point at the bright spot in the middle. Such a rare sight is a silent gift for anyone wise enough, like this cloudspotter, Ross McLaughlin, to have chosen the window seat.

ABOVE Subparhelic circle, diffuse arc, and parry anti-solar arc, spotted by Ross McLaughlin on a flight over Oslo, Norway.

THE MOUNTAIN ON NORWAY'S TINY GODØYA island appears to be wearing a wig. We can confirm, however, that this is not in fact a mountain wig at all. It is an orographic cloud. That's a cloud that forms as a result of wind rising to pass over raised ground. When an air-stream encounters an obstacle such as this island, the air lifts to pass over it, expanding as it does so. The expansion causes the air to cool. If it contains enough moisture, the drop in temperature encourages some of its moisture to condense into tiny droplets, which we see as cloud. An orographic formation perched on top of a peak like this is known as a cap cloud. It should, of course, be called a mountain toupee.

ABOVE A cap cloud, spotted by Marcus Murphy over Godøya island, Norway.

We wandered to the pine-forest
That skirts the Ocean's foam;
The lightest wind was in its nest,
The tempest in its home.
The whispering waves were half asleep,
The clouds were gone to play,
And on the bosom of the deep
The smile of Heaven lay.

From "To Jane: The Recollection" (1822), by Percy Bysshe Shelley.

STRATOCUMULUS CLOUDS OFF THE WESTERN COAST of Namibia and Angola are patterned with criss-crossing undulations, caused by atmospheric gravity waves produced by overlapping air masses. Dry air flowed westward over the Atlantic Ocean having been cooled through the chilly night of the Namib desert. Cool air is denser than warm, so it lifted the warmer, moister air out over the ocean, burrowing beneath the less dense warmer air. Shifting boundaries between two air masses are perfect conditions for atmospheric waves to develop. They are aerial versions of water waves but forming in the ocean of air that is our atmosphere. As they fan out over the Atlantic, they are revealed by the clouds. Wave upon wave. Ocean upon ocean.

ABOVE Stratocumulus undulatus produced by gravity waves, spotted off the coast of Africa by NASA's Terra satellite.

OPPOSITE A circumzenithal arc, spotted by Darya Light as sunlight refracts through the crystals of Cirrus clouds over Alameda, California, US.

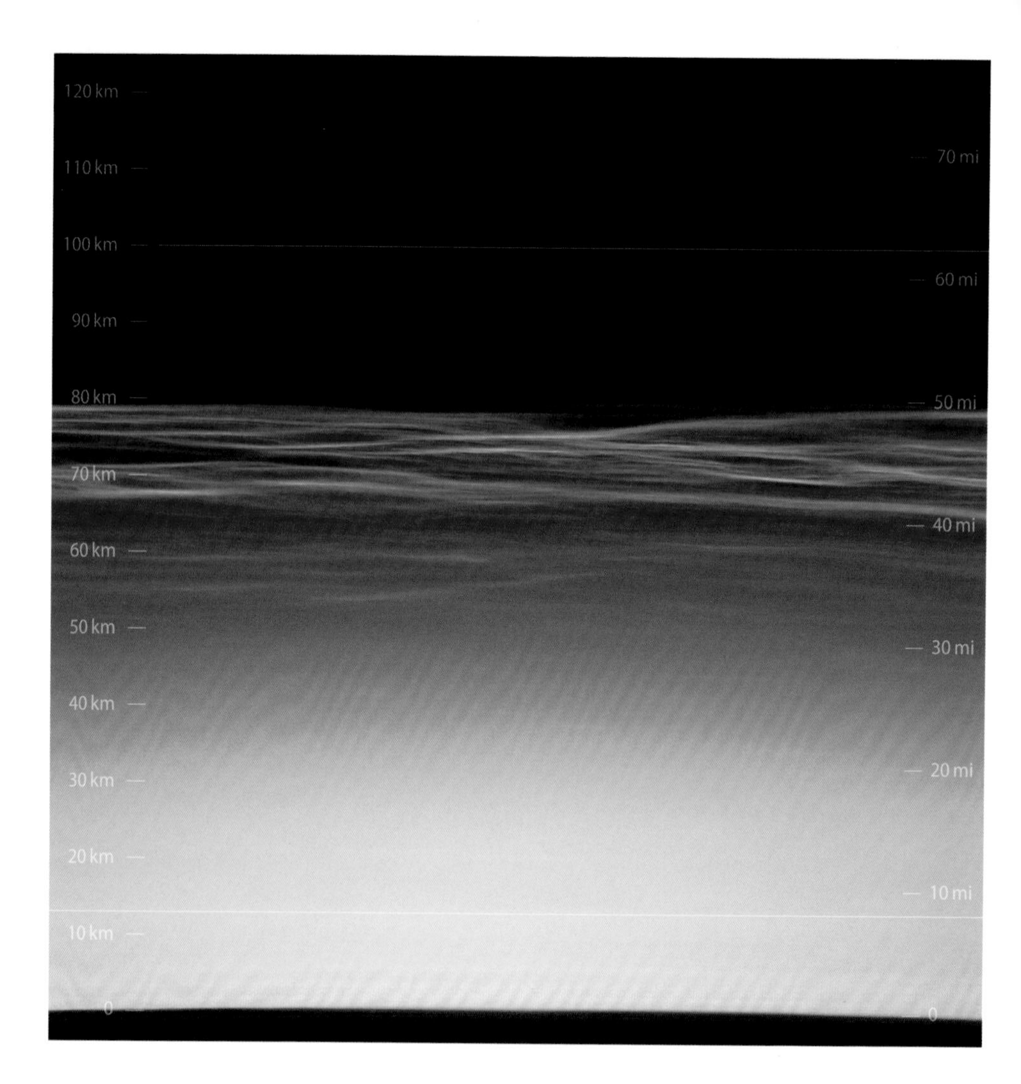

NOCTILUCENT CLOUDS, also known as polar mesospheric clouds, are the highest of all cloud types. The bottom line here shows the average altitude of the tropopause, below which all the familiar weather clouds form. The top one is where, by convention, we consider space to begin. Noctilucent clouds are only visible from latitudes of 50–70 degrees when the Sun is below the horizon so that they catch the light against a darkened sky. Delicate blue, gauze-like strands of ice, these clouds form in the coldest and driest part of our atmosphere. They are clouds at the fringes of space.

SOME CALL THESE BEAMS OF LIGHT "Jacob's Ladder" after the Bible's book of *Genesis* in which Jacob dreams of a ladder connecting Earth to Heaven. In Sri Lanka they are known as "Buddha's rays." Hawaiians call them "the ropes of Maui." The scientific name is crepuscular rays. Whatever they're called, they are our atmosphere's most sublime expression of light and shade.

ABOVE Crepuscular rays, spotted by John Findlay (Member 11,647) over Arnisdale, Highland, Scotland.

OPPOSITE Noctilucent clouds, spotted from the International Space Station by astronaut Jeff Williams, with altitude annotations added.

Till taught by pain, Men really know not what good water's worth.

From *Don Juan* (1819), by Lord Byron.

ABOVE The base of a brooding Cumulonimbus storm cloud, spotted over Toronto, Canada by Christina Connell (Member 43,390).

DARK PATCHES IN A LAYER OF CLOUD normally appear where the cloud is thicker and so lets less light shine through. Just sometimes, however, they are the shadows of higher clouds that are hidden from view. Such optics were revealed in this formation spotted by Randolph Harris. The wavelike ridges of Altocumulus undulatus clouds were casting shadows onto a shallow layer of Altostratus below, appearing from beneath like silhouettes on a Japanese paper partition.

ABOVE Altocumulus undulatus casting shadows on Altostratus, spotted by Randolph Harris (Member 45,004) over Church Creek, Maryland, US.

THIS IMAGE FROM THE US National Oceanic and Atmospheric Administration's GOES satellite system shows a curiously distinct line of clouds stretching across the Pacific Ocean. The cloud band marks the "intertropical convergence zone." Running roughly parallel to the equator, this is where the northeast and southeast trade winds come together, as air is drawn into the equatorial zone where the surface is warmed most directly by the Sun. The warmth creates huge thermals of rising air, drawing in winds from surrounding regions. At times this convergence can result in powerful storms, as are likely beneath the bands of cloud shown here. At other times the lack of any prevailing wind direction in the convergence zone means dead-calm seas, which is why the region has long been known by mariners as the "doldrums."

***PAREIDOLIA* (NOUN), FROM THE GREEK** words *para* ("beside, alongside, instead [of]"—meaning something faulty or wrong) and *eidōlon* ("image, form, shape"): the tendency to perceive a specific, often meaningful image in a random or ambiguous visual pattern.

ABOVE Just a random, ambiguous visual pattern, spotted by Linda Eve Diamond (Member 46,720) over Port Orange, Florida, US.

OPPOSITE A distinct line of storm clouds stretching across the Pacific marking the Intertropical Convergence Zone. Spotted by the Geostationary Operational Environmental Satellite system (GOES).

The happiest heart that ever beat
Was in some quiet breast
That found the common daylight sweet,
And left to Heaven the rest.

From "The Happiest Heart" (1894), by John Vance Cheney. First published in *Harper's New Monthly Magazine*, October 1894.

ABOVE A Cumulus heart, spotted by Jenny Shanahan on a flight from Chicago, Illinois, to Austin, Texas, US.

THE SPOTS, RINGS, AND ARCS OF LIGHT known collectively as halo phenomena are all caused by the sunlight shining through ice-crystal clouds. When the cloud's crystals are optically clear (like glass) and regular in shape, they act as prisms that refract the sunlight as it shines through, bending it at the point of passing into and out of the ice. The momentary glints from millions upon millions of ice crystals tumbling through large areas of the sky combine to appear as a halo phenomenon. And the most distinctive has to be the 22-degree halo. It is a bright ring, often with a reddish inner edge, centered on the Sun and with a radius of 22 degrees. That's the angle between the Sun and the halo edge, and it is equivalent to the distance between your thumb and little finger, when stretched out as wide as possible and held up at arm's length.

ABOVE A 22-degree halo produced by Cirrostratus cloud, spotted over Knox, Victoria, Australia, by Nicole Bates (Member 38,201).

WHEN THE BUDDHA wanted to cross the river Ganges near the ancient Indian city of Vaishali, the ferryman refused to take him over for free. The Buddha didn't do money; he managed without it, so instead he summoned a passing cloud to carry him over. Smart move.

ABOVE Detail from *Bouddha traverse le Gange*, illustration from "Vie illustrée du Bouddha Çakyamouni" in *Recherches sur les Superstitions en Chine* (1929), Vol. 15, by Henri Doré.

OPPOSITE, LEFT TO RIGHT Stratocumulus, spotted by David Rudas (Member 38,767). Altocumulus, spotted by Jiji Shedd (Member 42,40). Cirrocumulus, spotted by James Morrison (Member 41,584).

STRATOCUMULUS, ALTOCUMULUS, AND CIRROCUMULUS are classifications for clumpy layers of cloud forming at low, middle, and high levels of the troposphere, the region of our atmosphere where weather happens. They can be distinguished by the relative size of their clumps, or cloudlets. Stratocumulus, the lowest of the three, has clumps that appear largest because they are nearest. Cirrocumulus, the highest, has tiny clumps, which look more like grains on account of their distance. The mid-level Altocumulus clumps are somewhere between the two. These clouds can also be distinguished by how regular their clumps appear. The low Stratocumulus look more chaotic and disorderly because wind near the ground flows in a more turbulent manner due to its interaction with thermals and terrain. The clumps get more refined with altitude.

A SUNSET BECOMES ALL THE MORE FIERY with the appearance of a red rainbow. Formed when the Sun is low on the horizon, this is a rainbow colored not with the full light spectrum of day, but instead just the warm reddish hue of morning or evening. It is the result of a rainshower having just a single thread with which to weave its bow.

ABOVE Red rainbow, spotted over Mount Maunganui, North Island, New Zealand, by Graeme Blissett (Member 45,059).

ABOVE A towering Cumulus, also known as Cumulus congestus, rose high enough from the estuaries of Andros island in the Bahamas to grab the attention of a cloudspotting astronaut passing over far above aboard the International Space Station.

PAUL HENRY WAS AN IRISH LANDSCAPE PAINTER who attained great popularity during the 1920s and 1930s for his nostalgic portrayal of the countryside of Western Ireland. He reveled in light, land, and sky as he responded to his country's wild, open panoramas. In this painting vast Cumulus congestus clouds dominate the canvas, dwarfing the tiny sunlit cottages below. For Paul Henry what is above the line of the horizon is just as important, if not more so, than what's below it.

ABOVE Detail from *The Village by the Lake* (1929), by Paul Henry. Spotted by Mary Lazarus (Member 23,844).

The breeze at dawn has secrets to tell you. Don't go back to sleep.

From a quatrain by Jalaluddin Rumi, the 13th-century Persian poet and Sufi mystic, translated by Coleman Barks.

ABOVE Sunrise of Altocumulus, spotted over Gent, Belgium, by Frits Kuitenbrouwer (Member 13,684).

CUMULUS GRANITUS IS PILOT TERMINOLOGY for the peaks of snow-clad mountain that hide among the regular Cumulus clouds. It is a dangerous formation for low-flying aircraft.

ABOVE Mont Blanc hiding in Cumulus on the French-Italian border, spotted from the cockpit by Peter Leenen (Member 32,762).

OPPOSITE Someone seems to be sweeping the Cirrus clouds over Azalea Park, Florida, US, as spotted by Robyn Molnar.

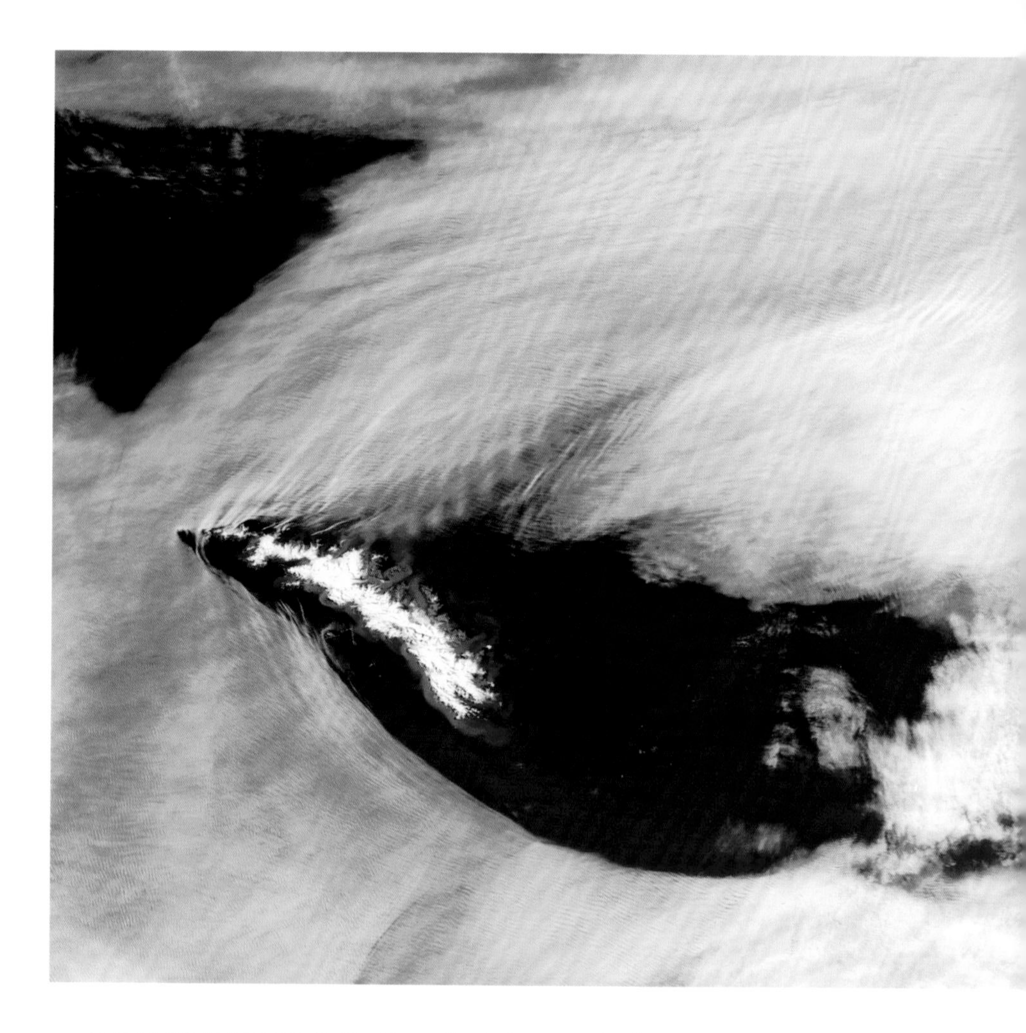

ONE OF THE TINY WILLIS ISLANDS, situated next to South Georgia Island in the midst of the Atlantic Ocean, parts the Altocumulus and Altostratus clouds. The mid-level clouds were likely composed of supercooled water droplets, and the cooling of the air as it lifted to pass over the island was probably enough to cause their droplets to freeze. If so, the ice crystals might have dissipated away as they fell into the warmer air below, leaving a widening gap in the island's wake. Whatever the mechanism, it is the best way for a tiny remote island to make its presence known out in the vast southern Atlantic.

ABOVE An island's wake in the clouds, spotted by Jeff Schmaltz using data from NASA's Terra satellite.

THESE ANTI-CREPUSCULAR RAYS are shadows cast by clouds in front of the Sun that is low on the horizon and shining from behind you. The beams of light and shade appear to converge as they recede off towards the anti-solar point, which is directly opposite the position of the Sun. Atmospheric haze such as moisture or dust reveals the shadows as it scatters the light. The convergence of anti-crepuscular rays towards the horizon away from the Sun is simply a trick of perspective. The beams of shadow and light are in fact as good as parallel.

ABOVE Anti-crepuscular rays, spotted near Taos, New Mexico by Heather Prince (Member 13,545).

SOMETIMES less is more.

ABOVE Cumulus humilis, spotted over Phoenix, Arizona, US, by Laura Simms (Member 32,141).

OPPOSITE Glass-plate photomicrographs of snow crystals by Wilson Bentley.

WILSON BENTLEY, FROM JERICHO, Vermont, US, photographed his first snowflake in January 1885. He had been fascinated by the varied appearances of atmospheric ice crystals since he first observed one through a magnifying glass. Bentley perfected a technique for collecting and transferring them unharmed to glass slides in order to capture them on film through a microscope. He photographed over 5,000 intricate crystals and became known as "The Snowflake Man." He died in 1931, aged 66, of pneumonia after walking home to his farm through a blizzard. "Under the microscope," Bentley had written in 1925, "I found that snowflakes were miracles of beauty . . . a masterpiece of design and no one design was ever repeated. When a snowflake melted, that design was forever lost. Just that much beauty was gone, without leaving any record behind."

HAS THE MOUNT TARANAKI VOLCANO in New Zealand just had an unfortunate scrape with a passing Altostratus cloud? No, the gash in the cloud layer is actually a distrail, short for dissipation trail, created by an aircraft from nearby New Plymouth Airport climbing through the layer of cloud. The cooling effect of turbulence around the plane's wings was enough to encourage the supercooled droplets that made up the cloud layer to freeze into ice crystals. As the crystals grew, they fell from the cloud, evaporating away in the warmer, drier air below to leave a gap in the Altostratus. From where Graham Billinghurst spotted this distrail, Mount Taranaki's peak just happened to be perfectly aligned with its end. Phew—no need to exchange insurance details.

ABOVE Aircraft dissipation trail over Mount Taranaki, North Island, New Zealand, spotted by Graham Billinghurst (Member 24,513).

THE HOPI TRIBAL GROUPS of what is now Arizona, US, used to undertake pilgrimages to the Grand Canyon, which they called *Ongtupqa*. On these journeys they would often stop to inscribe symbols onto the rocks at Echo Cliffs. Some of the carvings date back to the 1200s and the site, known as Tutuveni, is now a protected ancient monument with over 5,000 symbols left by different clans. One such clan carved along the bottom of a boulder this row of triangular shapes with lines beneath to depict raining clouds. As a result, the clan is now referred to by historians as the Cloud Clan. We like to think of them as an early chapter of the Cloud Appreciation Society.

ABOVE Cumulus congestus clouds over Tutuveni, Coconino Country, Arizona, US, spotted by the Cloud Clan and photographed beneath fair-weather Cumulus by Tom Bean (Member 41,135).

WHEN A STRANGE RING appeared in the sky over Warwickshire, England, cloudspotter James Tromans asked us why. It was time for some cloud detective work. There was something unnatural about its appearance, as if it were man-made. It occurred to us that the photograph was looking towards Coventry Airport. Might this cloud effect have been caused by a plane? Aircraft condensation trails can sometimes form these zip-like lobes, caused by the interaction of the vortices around the aircraft wings. Conclusion: The formation was caused by a plane just above the cloud base turning in a holding pattern as it waited to land. Its contrail was hidden within the cloud, with the lobes caused by its wing turbulence extending below. Another case solved by the Cloudspotting Detective Agency.

If lightning is the anger of the gods, then the gods are concerned mostly about trees.

Attributed to Lao-Tzu,
ancient Chinese philosopher from the 6th century BC.

ABOVE Lightning from a Cumulonimbus, spotted over Schleiz, Thuringia, Germany, by Juergen K. Klimpke (Member 22,868).

OPPOSITE Turbulence lobes, spotted by James Tromans beneath the condensation trail of an aircraft in a holding pattern over Hampton Lucy, Warwickshire, England.

ABOVE A breakdancer busts some moves over Zalau, Romania, as spotted by Fiorella Iacono (Member 9,702).

OPPOSITE A pride of huge Cumulonimbus clouds prowl across the Straits of Florida and the island of Cuba, as spotted by NASA astronaut Ricky Arnold aboard the International Space Station. Storm clouds like to hunt in packs like this.

IN 1942, WHEN THIS PAINTING by John Rogers Cox was first exhibited in New York's Metropolitan Museum of Art, few could have missed the symbolism of the dark Cumulus congestus clouds building on the horizon. A crossroads. Idyllic golden grain fields. A bank of darkening storm clouds ahead. Cox painted it shortly after America joined the Second World War.

ABOVE *Gray and Gold* (1942), by John Rogers Cox, spotted by Steven Grueber (Member 41,808).

OPPOSITE LEFT Upper tangent arcs, spotted over Maleč, Plzeňský, Czech Republic, by Karel Jezek (Member 34,987). **OPPOSITE RIGHT** Upper tangent arcs, spotted over Eiksmarka, Akershus, Norway, by Monica Nitteberg.

SOME HALO PHENOMENA, like the upper tangent arcs shown here, vary in shape depending on the elevation of the Sun. Halo phenomena are caused by sunlight refracting as it passes through tiny hexagonal ice crystals in the atmosphere, perhaps reflecting off their inner surfaces as it does so. Tangent arcs tend to form about once a month in the skies over temperate regions, so they are relatively common. When the Sun is low on the horizon, seen above right, where is is probably at about 10 degrees, the tangent arc is shaped like a "V." As the Sun climbs higher in the sky, seen above left, where it is more like 30 degrees, the arc is much more flattened, its edges curving around the Sun. Tangent arcs can be seen below the Sun when viewing from an elevated position like a mountainside or an aircraft.

ONE REASON CIRROCUMULUS is the least common of the ten main cloud types is that it never hangs around for long; it is a cloud that is always in transition to another form. Its high elevation means that the droplets in its little cloudlets soon freeze into ice crystals. That is when it starts changing into one of the other high clouds: flowing into the long ice trails of Cirrus or diffusing into the subtle, crystal layer of Cirrostratus. We like to categorize clouds, to arrange them into neat groups like we do with the rest of nature. Cirrocumulus, the most fleeting and transient of the main types, reminds us that clouds always refuse to be boxed in.

CHIAROSCURO IS THE TECHNIQUE developed by artists in the Renaissance—Leonardo da Vinci, Carravagio, and Rembrandt in particular—that uses strong contrasts of light and dark tones to heighten the drama of a painting. Now where, we wonder, did they get that idea?

ABOVE Cumulus and Cumulonimbus, spotted at sunset off the coast of Pagedongan, Banten, Indonesia, by Nizma Arifin (Member 36,177).

OPPOSITE Cirrocumulus by Jorge Figueroa Erazo over Guatemala City, Guatemala.

A BROAD RIVER OF DUST is connecting the Sahara in western Africa at the bottom of this image with the Amazon basin in South America at the top. While it can have a negative impact on air quality in the Americas, the Saharan dust cloud is not as unwelcome a visitor as you might think. The vast quantities of particulates bring to the Amazon rainforest a much-needed dusting of fertilizer. The airborne phosphorus from the Sahara's ancient seabed arrives to replace the nutrients that the abundant rains of the region leach from the soils and carry down the mighty river out to sea.

ABOVE Composite photograph assembled from data acquired by the Suomi NPP satellite operated by NASA.

ABOVE "Falling stars as observed from the balloon," an illustration from *Travels in the Air* (1871), by the Victorian pioneering balloonists James Glaisher, Camille Flammarion, Wilfrid de Fonvielle, and Gaston Tissandier.

ABOVE A floret of broccoli, out jogging over North Miami Beach, Florida, US. Spotted by Adam Littell.

Between two stupendous mountains of the low stratum under the evening red, clothed in slightly rosaceous amber light, through a magnificent gorge, far, far away, as perchance may occur in pictures of the Spanish coast viewed from the Mediterranean, I see a city, the eternal city of the west, the phantom city, in whose streets no traveller has trod, over whose pavements the horses of the sun have already hurried, some Salamanca of the imagination.

From the American naturalist Henry David Thoreau's journal on July 10, 1851.

ABOVE Crepuscular rays, spotted by Mark Hayden over Butte, Montana, US.

ABOVE A formation of Altocumulus lenticularis, bathed in the glow of sunset, mimics the contours and strata of the Grand Canyon below. Spotted over Arizona, US, by John Bigelow Taylor (Member 42,972).

OPPOSITE A halo display in diamond dust, spotted by Joost van Ekeris over Stadle, St. Anton am Arlberg, Tyrol, Austria.

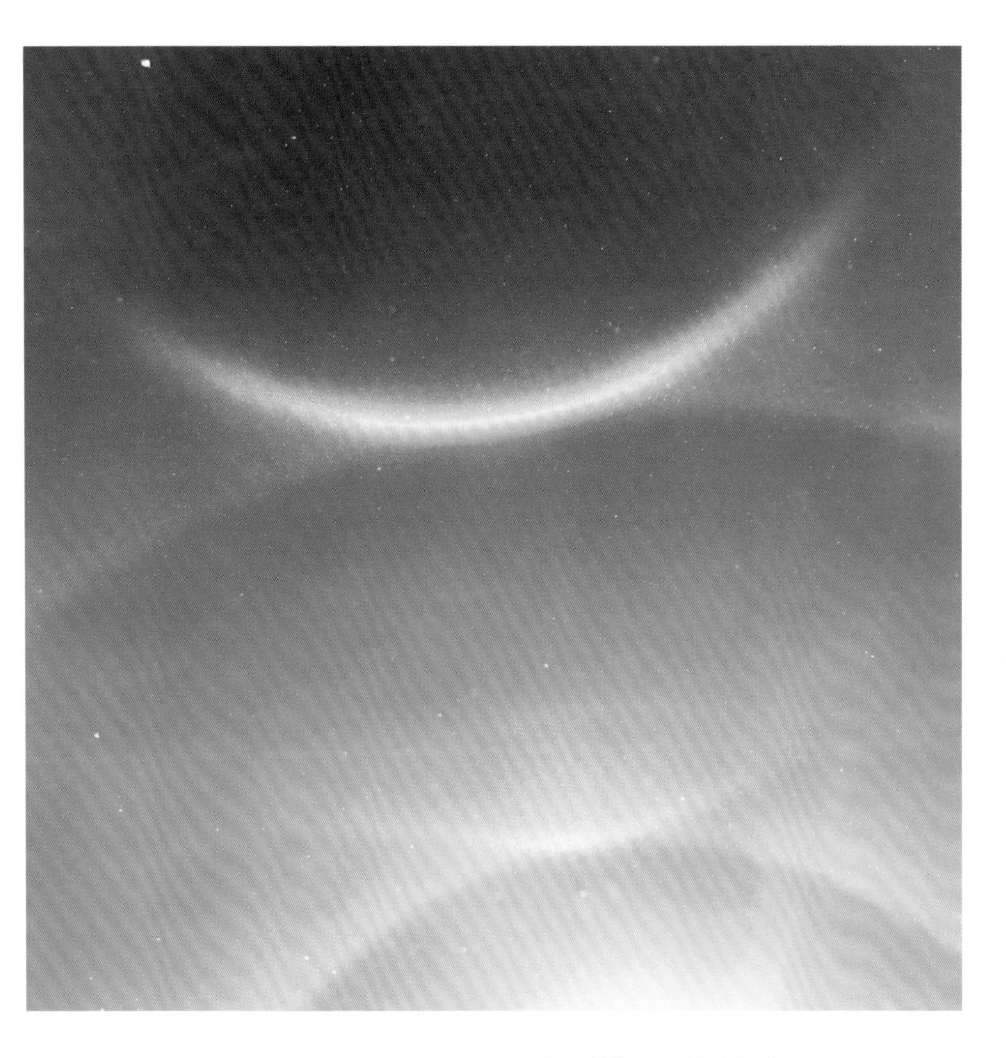

THE EMINENT HARVARD BIOLOGIST E. O. Wilson said, "The first step to wisdom . . . is getting things by their right names." This display of optical phenomena is a case in point. Caused by the sunlight shining through the sparkling ice-crystal fog known as diamond dust, these arcs of white and colored light each have a name. From top to bottom the halos include a circumzenithal arc, a supralateral arc, a faint parry arc, an upper tangent arc, and a 22-degree halo. The position of the Sun is off to the bottom right of the image. Just remember, though naming may be the first step towards wisdom, you should never lose touch with your ability to gaze up in silence at the intricate beauty of our atmosphere.

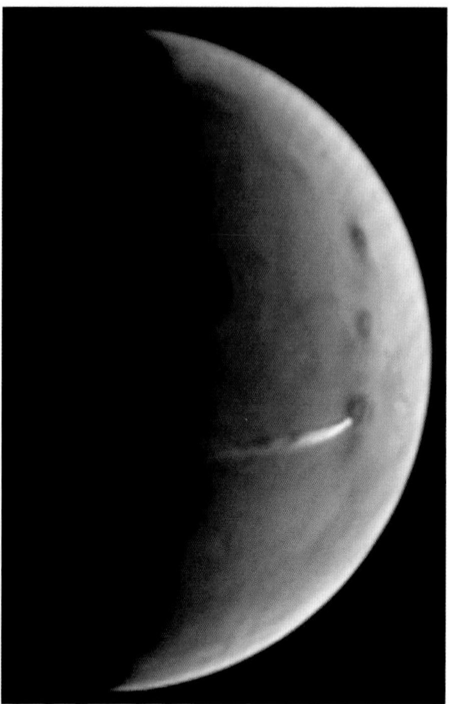

RIGHT A long cloud trails from the huge Arsia Mons volcano on Mars. Water ice clouds like this form regularly as winds flow over the volcano, tending to lengthen downwind of the peak as the Martian day progresses. Such a formation would be known here on Earth as a banner cloud. The Martians, no doubt, have their own name for it.

BELOW Moisture rising from the cooling towers of power stations can produce clouds when air temperatures are cool enough. Spotted by Raymond Kenward over the Isle of Grain, Kent, England, these clouds of heavy industry are known officially as Cumulus homogenitus, meaning man-made.

OPPOSITE Cirrus cloud, spotted over Lake Ferry, South Wairarapa Coast, New Zealand, by Kym Druitt (Member 19,908).

IN SHAKESPEARE'S PLAY *Hamlet* the fawning courtier Polonius sees in the clouds whatever he's told to see by Hamlet, Prince of Denmark.

> HAMLET: *Do you see yonder cloud that's almost in shape of a camel?*
>
> POLONIUS: *By the mass, and 'tis like a camel indeed.*
>
> HAMLET: *Methinks it is like a weasel.*
>
> POLONIUS: *It is backed like a weasel.*
>
> HAMLET: *Or like a whale.*
>
> POLONIUS: *Very like a whale.*

The lesson here is don't let anyone tell you what to see in the clouds. That said, wethinks this one is shaped rather like a sperm whale.

THIS ILLUSTRATION WAS MADE BY Myles Birket Foster to accompany a publication of Henry Wadsworth Longfellow's 1841 poem "The Rainy Day," which contains these famous lines:

Be still, sad heart! and cease repining;

Behind the clouds is the sun still shining;

Thy fate is the common fate of all,

Into each life some rain must fall.

ABOVE Pen and ink illustration (1850s) for Longfellow's "The Rainy Day" by Myles Birket Foster.

ABOVE Earthrise on July 20, 1969, as viewed from the Moon by the astronauts on the Apollo 11 mission, shortly before separation of the Lunar Module from the Command Module.

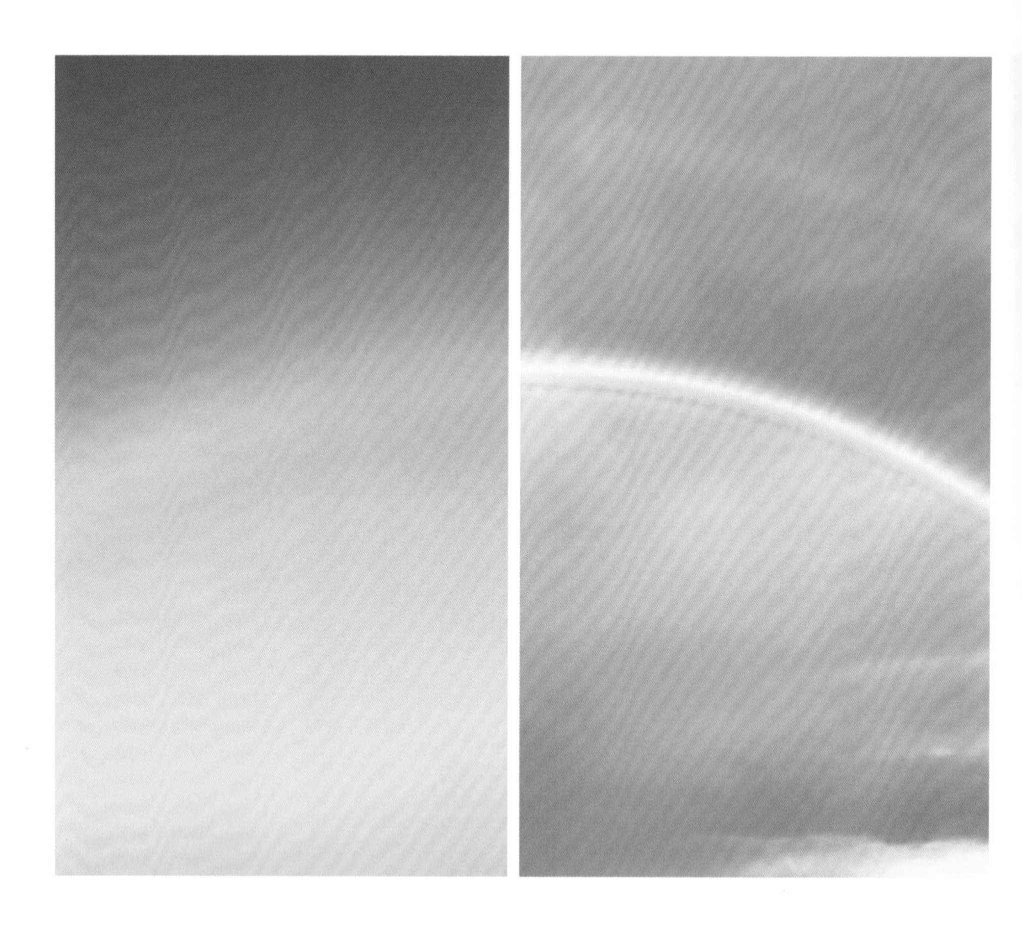

A CLOUDBOW IS THE PASTY-FACED COUSIN of a rainbow. Both optical effects are caused by droplets, and both appear when the Sun is fairly low in the sky and shining from behind you. A rainbow gets its healthy, bright coloration from the larger droplets of rain, which separate the wavelengths of light by refraction as they reflect it back at you. A cloudbow does much the same, but since the cloud droplets are far smaller, and so closer in size to the wavelengths of light, they have the effect of blurring the colors together more, bleaching them out. Compared to those of its radiant cousin—the bow with all the fame and glory—a cloudbow's hues are so weak, so frail, as to be barely distinguishable. Sometimes, it can muster no color at all.

ABOVE LEFT A cloudbow from fog, spotted over Miami-Dade County, Florida, US, by Chase Vessels (Member 40,104).

ABOVE RIGHT A rainbow, spotted over Park County, Colorado, US, by Leslie Cruz (Member 41,661).

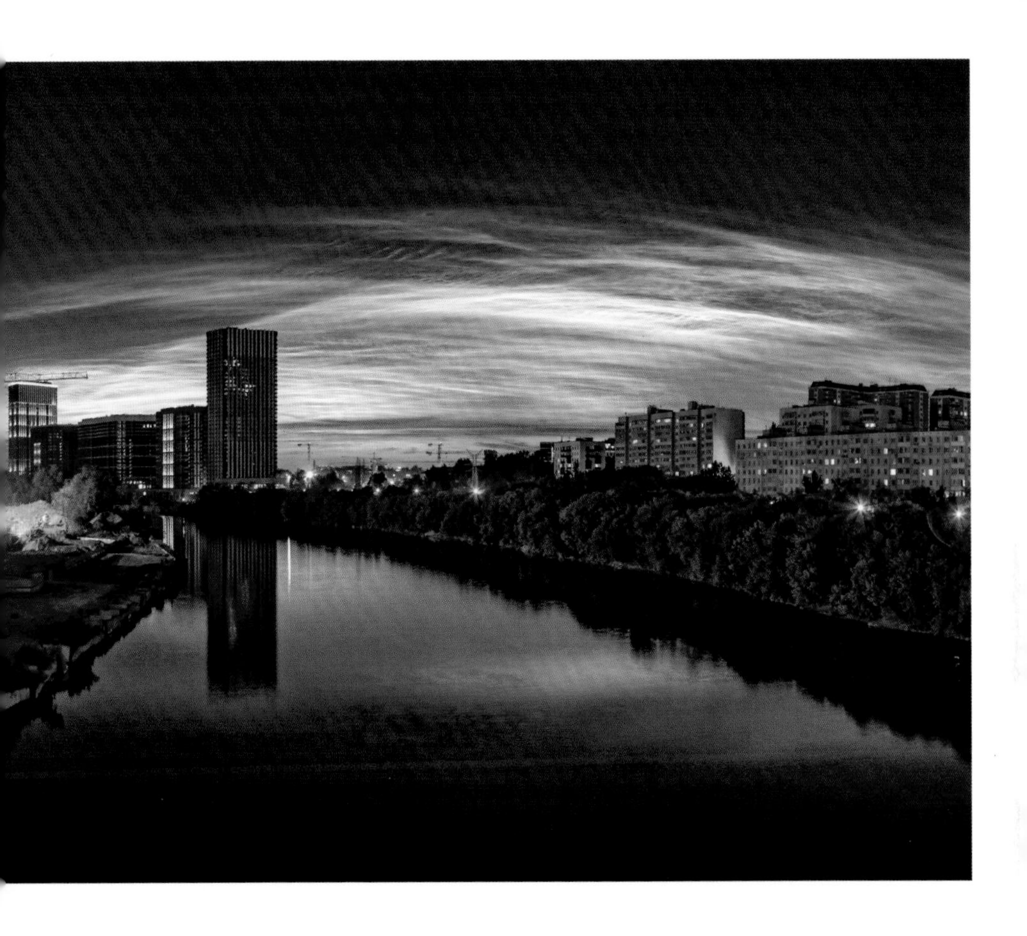

NOCTILUCENT CLOUDS are only visible from higher latitudes, between 50 degrees and 70 degrees, and after dark, when they shine out against the night sky as ghostly, bluish ripples. They form at around 85 kilometers (53 miles), which is right up in the mesosphere. Being so high, they still catch the light long after the Sun has dipped below the horizon and the lower atmosphere is in shadow. This is why their name comes from the Latin for "night-shining." The season for spotting noctilucent clouds is in the middle of summer. This is because, counterintuitively, warmer temperatures in the lower atmosphere coincide with colder temperatures up in the mesosphere, which are needed for the subtle ice clouds to form.

ABOVE Noctilucent clouds, spotted by Dmitry Kolesnikov over the Moskva River, Fili, Moscow, Russia.

YOU WOULDN'T THINK of the Sahara desert as a place to go sledding, but that's exactly what residents of the northern Algerian town of Aïn Séfra did there in January 2018. Up to 30 centimeters (12 inches) of snow fell on the slopes of the highest sand dunes near the town. Snowfalls are not in fact unprecedented in the Sahara, where night-time temperatures often dip below 0°C (32°F). The previous time it snowed in the area was in December 2016. Then, as on this occasion, the snow lasted no longer than a few hours before melting away in the desert heat of the day.

ABOVE Snow in the Sahara, spotted by Joshua Stevens, combining data from NASA's Landsat 8 satellite and Shuttle Radar Topography Mission.

DISC-SHAPED LENTICULARIS CLOUDS occasionally appear stacked one above the other like this. It is a formation known by the French term *pile d'assiettes*. The formation can appear in the lee of hills or mountains when the upper winds consist of alternating drier and moister layers of air. The rising and dipping mountain waves that develop in the flow of air can cause plates of lenticularis clouds to form within the moister layers with gaps in the drier ones between.

ABOVE Altocumulus lenticularis clouds, spotted by Laura Stephens over Rothesay in Bute, Scotland.

THIS IS THE CLOUD supplementary feature known as mamma. The distinctive pouches of cloud can appear on the underside of several of the main cloud types, but the most dramatic ones are to be found beneath the high canopies that spread out at the top of large Cumulonimbus storm clouds. Since they tend to appear towards the rear of the cloud's direction of movement, a sky full of dramatic mamma is usually an indication that a storm, though near, is heading away from you.

ABOVE Cumulonimbus with mamma, spotted over Boydston, Texas, US, by Christina Brookes (Member 33,764).

OPPOSITE A Cirrus bird of prey looks for easy pickings on the blue run, spotted by James McAllister over Les Allues, Auvergne-Rhône-Alpes, France.

THIS LONG, HORIZONTAL TUBE OF CLOUD was spotted by Ko van Hespen ahead of a developing Cumulonimbus along the coast of the Netherlands. It is known as a volutus, or roll cloud. One that forms ahead of a storm like this one did is closely related to the feature known as an arcus, or shelf cloud, which has a similar roll-like appearance but is attached to the main storm rather than separated from it like this. A storm-induced volutus like this can travel, rolling along at speeds of up to 50 kilometers per hour (30 miles per hour), within an invisible wave of air pushing out ahead of the storm.

ABOVE Volutus, or roll cloud, ahead of a developing Cumulonimbus, spotted over Zandvoort, Netherlands, by Ko van Hespen (Member 36,654).

OPPOSITE *Equivalents* (1926), gelatin silver print, by Alfred Stieglitz. Suggested by Lou Morgan (Member 28,857).

THE AMERICAN PHOTOGRAPHER Alfred Stieglitz was the first to explore the potential of photography as an abstract art, and he did so by turning his camera to the sky. From 1923 to 1934, he made around 220 photographs of clouds, most of which had no land or objects to act as reference points, focusing simply on the abstract celestial forms. They feel natural for a modern-day cloudspotter, but they were revolutionary in the 1920s. Stieglitz named the ongoing series *Equivalents*. He considered them to be equivalent to feelings, to states of mind. They were highly influential in the development of photography as an art form. The American landscape photographer Ansel Adams said in 1948 that his first "intense experience in photography" was seeing *Equivalents*.

ABOVE Cumulonimbus storm clouds embody the sublime power of our atmosphere. They are the monumental architecture of air. This one was spotted over Huntersville, North Carolina, US, by Lauren Antanaitis (Member 25,124).

OPPOSITE A cross-section of a large hailstone, spotted in *L'Atmosphere* (1872) by Camille Flammarion.

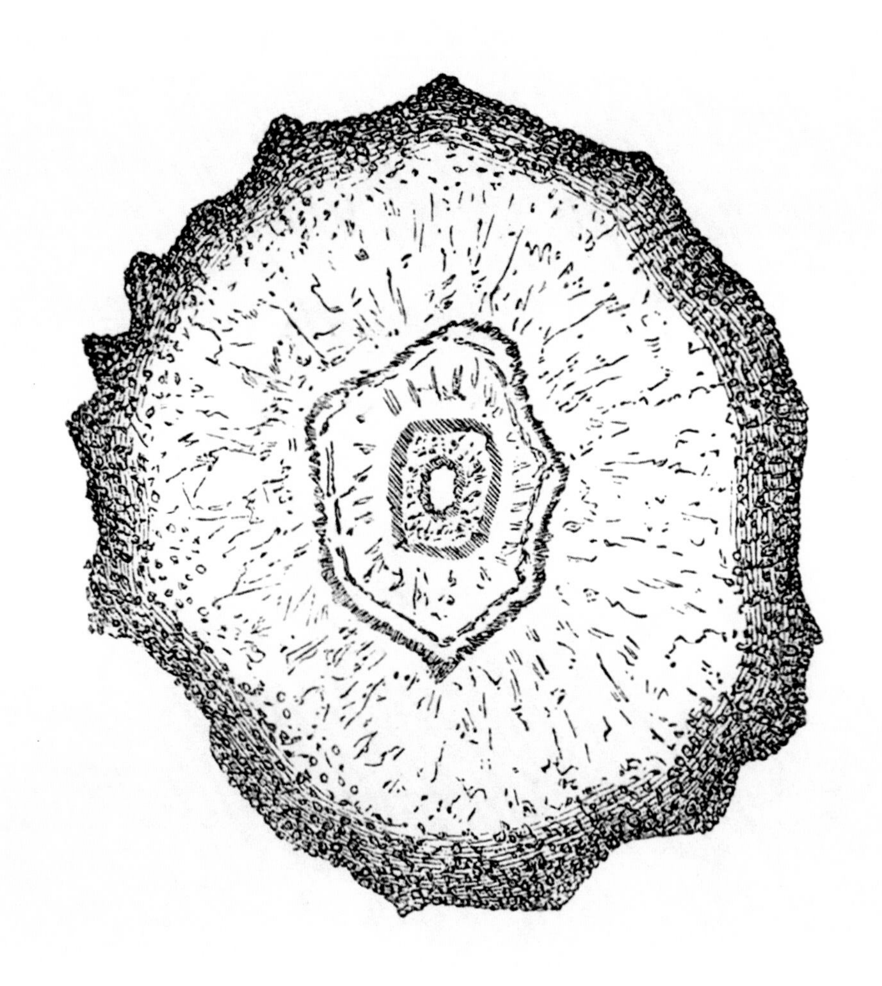

THIS ILLUSTRATION APPEARED in an 1872 book called *L'Atmosphere* by Camille Flammarion. It depicts the cross-section of a large hailstone, showing frozen bubbles radiating out from the stone's center, as well as a distinctive pattern of concentric rings. These inner rings are typical of hailstones. They reveal how ice builds inside a large Cumulonimbus storm cloud. The hailstones grow gradually as they rise and fall, swept up and down by the violent air currents within the cloud. Each stone is coated in water as it falls through rain in the cloud's lower regions. This soon freezes as the hailstone is swept back up by huge updrafts into the cloud's icy upper reaches. One air current feeds into the other, so hailstones rise and fall like this over and over again. They are nature's icy gobstoppers, made layer upon layer within the turbulent belly of the Mother of All Clouds.

As distant prospects please us, but when near

We find but desert rocks and fleeting air.

From *The Dispensary* (1699), Canto III, by Samuel Garth.

ABOVE Fog, spotted by Lodewijk Delaere over Mount Batur, Bali, Indonesia.

OPPOSITE A green flash, spotted over the Pacific from Albion, California, US, by Dennis Olson (Member 28,231).

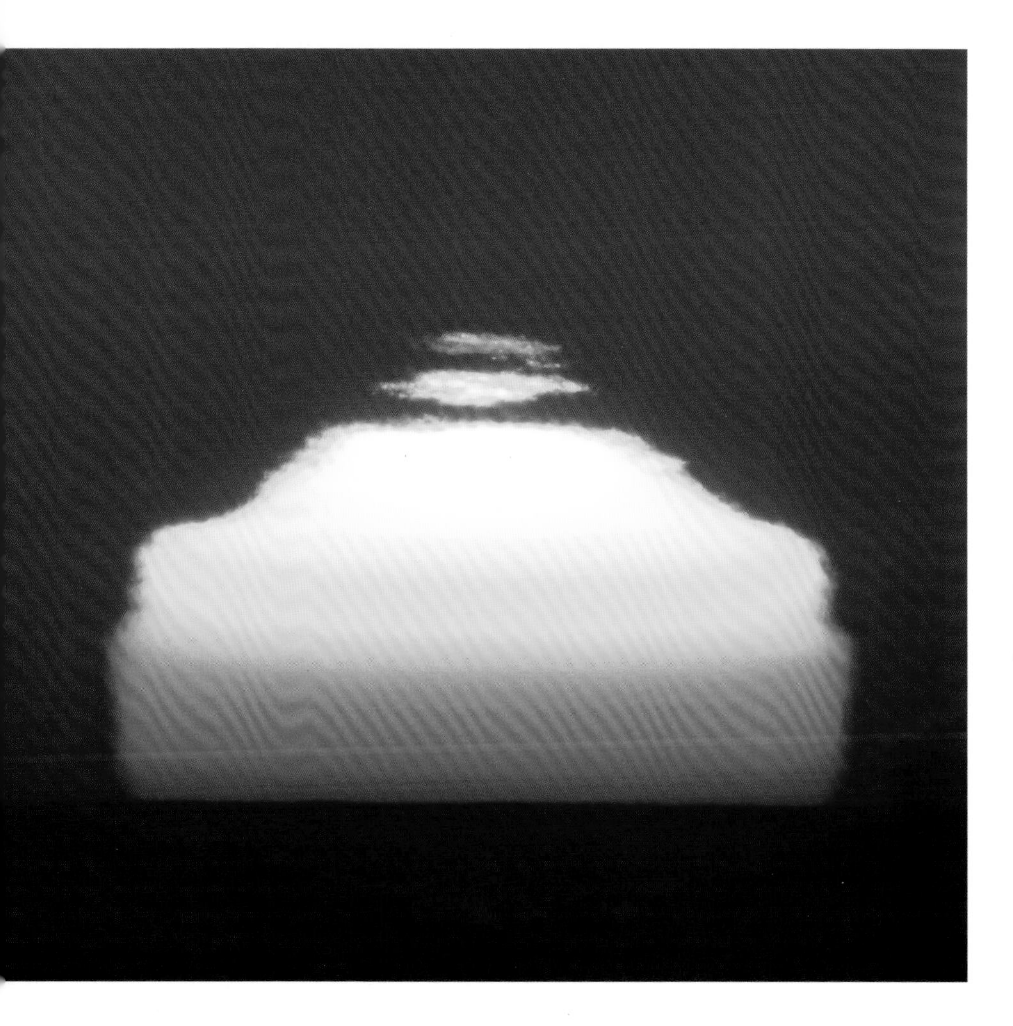

AT THE TOP OF THIS IMAGE of the setting Sun, a small disc of light is glowing vivid emerald. This is the optical effect known as a green flash. The flicker of color above the Sun on the horizon lasts only a few seconds and depends on particular atmospheric conditions. The Sun's rays are refracted upwards by the Earth's atmosphere. The green part of the spectrum refracts more strongly than the other colors of the low Sun, causing it to appear to have a green rim along its upper edge. The green flash can appear separated off into a disc of light like this when air temperatures near the surface cause the effects known as mock mirages. These are sometimes seen looking out to sea when ocean currents cool the low air to create an abrupt temperature inversion.

ABOVE Good morning, Titan. Daylight breaks over the north pole of Saturn's largest moon, and a milky haze in the moon's upper atmosphere is spotted by NASA's Cassini spacecraft. The misty layers are formed not of water but of complex molecules that originate from the atmosphere's methane and nitrogen gases.

OPPOSITE Devil hidden in the clouds of Giotto's 13th-century fresco, spotted by historian Dr. Chiara Frugoni.

THE HISTORIAN CHIARA FRUGONI had been studying the frescos of Giotto for over 30 years when, in 2011, she spotted something no one had noticed before in one of his paintings in the Basilica of St. Francis in Assisi, Italy. There in the fresco, hidden in the corner of a cloud, Dr. Frugoni could just make out the shape of a face. It appeared to be that of a devil, complete with horns. The fresco had been admired by countless visitors over the 720 or so years since Giotto painted it. But it seems not one had ever noticed the face. Frugoni's find threw up many questions. Why would the artist hide a devil in the clouds? Did he do so with or without the knowledge of the Franciscan friars who commissioned the work? One thing we do know for sure: This 13th-century fresco is the earliest known example in Western art of a shape depicted in the clouds.

WHEN THE SKY IS CLEAR, moonless, and away from light pollution, either an hour before dawn or after sunset, you may notice a pale, cone-shaped glow rising from the horizon in the direction of the hidden Sun. This is called the zodiacal light and originates from far out in our solar system. It is sunlight that is being reflected back to Earth by billions of minute interstellar dust particles that orbit the Sun as far out as the planet Jupiter. The light follows the line of the zodiac, the path along which the Sun appears to travel through the sky from Earth, as that is the plane that the dust lies in. In exceptionally dark and clear skies the light can be seen to span the whole sky, but usually it just forms this patch, which in the morning is also known as a false dawn.

ABOVE Zodiacal light, spotted from the European Southern Observatory's La Silla Observatory, Chile, by Yuri Beletski.

I found every breath of air, and every scent, and every flower and leaf and blade of grass, and every passing cloud, and everything in nature, more beautiful and wonderful to me than I had ever found it yet.

Esther Summerson ventures outside for the first time after a long period of illness, from *Bleak House* (1853), by Charles Dickens.

ABOVE Cumulus humilis, spotted by Margot Redwood over Eastwood, West Yorkshire, England.

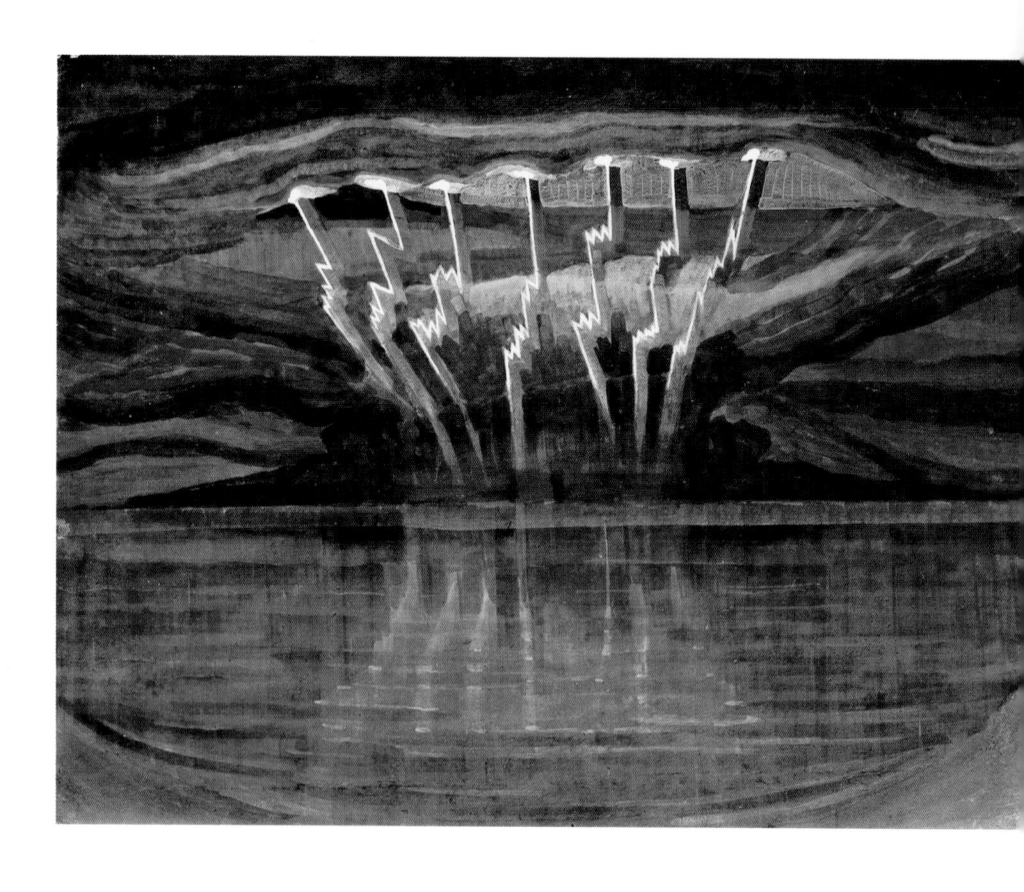

ABOVE *Lightning* (1909), by the Lithuanian pioneer of abstract art Mikalojus Čiurlionis.

STRATOCUMULUS UNDULATUS RADIATUS clouds at night over southern England. The undulations of air that give rise to these low rolls of cloud result from the combination of shearing winds—when speed increases markedly with altitude—and stable conditions. This sets up a rising and dipping of the air that, when conditions are right, causes cloud to form in the rising parts and gaps of open sky to clear in between. Seeming to fan out due to perspective, the rolls drift silently through the calm night air, largely unnoticed by all those sleeping below.

ABOVE Stratocumulus undulatus radiatus clouds, spotted by Chris Damant over Swanbourne, Buckinghamshire, England.

ADVECTION FOG CAN FORM when moist air is cooled as it drifts gently over a cold surface. The surface might be cool ground or, as here in the San Francisco Bay, chilly coastal waters where ocean currents well up from the deep. A gentle breeze drifting into the bay will have picked up moisture out over the warmer open waters of the Pacific. This can condense as the air cools over the bay, forming into the tiny droplets that appear as surface-level Stratus, or fog. Summer conditions in San Francisco are perfect for advection fog, so it greets the Golden Gate Bridge with the embrace of old friends.

ABOVE Advection fog, spotted on the Golden Gate Bridge, San Francisco, California, US, by Michael Warren (Member 37,489).

The voice of Nature loudly cries,

And many a message from the skies,

That something in us never dies . . .

From "New Year's Day: to Mrs Dunlop" (1790), by Robert Burns.

ABOVE Altocumulus with crepuscular rays, spotted over New Deer, Aberdeenshire, Scotland, by Roger Lewis (Member 36,182).

OF ALL THE CLOUD TYPES, Cumulus are the best impersonators. They are the most solid-looking clouds with the clearest outlines, and so the best lures to our imaginations. We can't help but interpret these three-dimensional stacks of moisture as familiar objects or faces. Here's one that's both: the stone bust of Queen Nefertiti, who reigned over Egypt from c. 1353 to 1336 BC. The bust is admired by thousands of visitors each year in the Neues Museum in Berlin, Germany. The Cumulus equivalent was admired by Paula Maxwell alone. She watched it float across the blue, lasting for no more than a moment.

ABOVE Queen Nefertiti, spotted by Paula Maxwell over Santa Clarita, California, US.

SUN DOGS, also known as parhelia or mock suns, appear on either side of the Sun as its light is refracted through hexagonal-shaped ice crystals in the atmosphere. This upper pair was spotted in the Cirrostratus clouds over Lake Maggiore, Italy, by Michela Murano and Valeriano Perteghella. The lower pair was spotted some 500 years before in the *Nuremberg Chronicle*. This Biblical history written by Hartmann Schedel and illustrated by the workshop of Michael Wolgemut was published in Nuremberg, Germany, in 1493, and is one of the first ever illustrated printed books. We are glad they chose to put something useful in it.

ABOVE Sun dogs formed by Cirrostratus, spotted over Lake Maggiore, Italy, by Valeriano Perteghella (Member 31,636).

BELOW Sun dogs, spotted in the *Nuremberg Chronicle*, 1493.

WHAT MAKES THIS relatively small interstellar cloud interesting is the bright boomerang shape towards its top. This is a newly formed star emerging from the cloud and leaving a wake in the gas and dust of the nebula. Since the sun is hurtling along at around 200,000 kilometers per hour (124,000 miles per hour), this particular boomerang will not be coming back.

ABOVE Nebula IRAS 05437+2502, spotted through the Hubble Space Telescope.

OPPOSITE The mid-level cloud Altocumulus generally produces the best sunsets. Here is one of its creations, spotted over Inle Lake, Myanmar, by Steven Grueber (Member 41,808).

ABOVE Parallel lines of Cirrus fibratus clouds dispersing into floccus formations resemble stalks of wheat arrayed across the December skies of Sierra de Álamos, Sonora, Mexico. Spotted by Suzanne Winckler (Member 41,844).

OPPOSITE Detail of *A Wanderer Above the Sea of Fog* (1817), by Caspar David Friedrich.

CASPAR DAVID FRIEDRICH'S 1817 painting *Wanderer Above the Sea of Fog* is a symphony of Stratus. The German Romantic painter composed the scene from sketches made on walks in the Elbe Sandstone Mountains of Saxony and Bohemia.

People love the feel of fog on their skin, immersed, wet and cold, but gentle and soothing. It's a primary experience.

Fujiko Nakaya, 86-year-old contemporary Japanese artist.

ABOVE Fog on the Shenipsit Trail, Glastonbury, Connecticut, US, spotted by Dennis Paul Himes (Member 5,003).

OPPOSITE Banner cloud, spotted by John Callender (Member 26,942) trailing from the Matterhorn, on the border between Switzerland and Italy.

FEW THINGS CAN IMPROVE a view of the majestic Matterhorn, which stands at 4,478 meters (14,690 feet) on the border of Switzerland and Italy. The right sort of cloud, however, might just be one of those things. This one was spotted by John Callender, unfurling on the leeward side of the mountain peak, where stiff winds caused the air pressure to drop. This low pressure sucked up air lower down the leeward slopes. As it rose, the air cooled—both because it expanded and because it mixed with the colder winds above—enough for moisture within to condense into droplets. The formation is known as a banner cloud, but you could just as easily think of it as an enormous windblown scarf, snagged on the very tip of the Matterhorn's peak.

IN THIS PHOTO, just to the left of the Moon, is a rarely captured phenomenon called a sprite. Its red streaks are huge, fleeting electrical discharges forming high up in the atmosphere, above the lightning strikes in a major thunderstorm below. The lightning below appears as a bright white patch near the surface. Sprites shoot up 50–90 kilometers (30–50 miles) above the storm. Unlike the extremely hot lightning, these red, jellyfish-like flashes are actually cold plasma discharges. Each lasts no more than a few milliseconds, so you need to spot them with a very high-speed camera.

ABOVE Red sprite, spotted over the US and Central America by astronauts aboard the International Space Station.

ABOVE A bird sings to the Sun with such a beautiful melody that it causes a sun dog to appear in the sky behind it. Spotted over Erbil, Iraqi Kurdistan, by Azhy Chato Hasan (Member 1,687), and also known as Cumulus below Cirrostratus, forming the ice-crystal optical effect known as a parhelion.

THE STILL, HIGH-PRESSURE WEATHER beloved by balloonists is associated with sinking air. This tends to suppress cloud formation, reduce winds, and encourage fog. Clear night skies mean the ground cools more rapidly than it would beneath the insulation of cloud cover. Without clouds up above, the ground radiates its warmth up into space faster, so it becomes colder through the night. By morning, the cold ground will have chilled the air drifting over it enough to encourage moisture to condense into droplets of radiation fog. This example heralds the beginning of the Australian summer, as the fog shrouds the open fields like a blanket, golden in the morning light. It will only last until the rising Sun has warmed the ground once more. Then the fog will disappear—and with it the balloonist's dreams of a nice, soft landing.

ABOVE Radiation fog, spotted by Phil Chapman over the Yarra Ranges in Victoria, Australia.

IN 2004, THE UNITED STATES POSTAL SERVICE issued a set of 15 $0.37 "Cloudscapes" stamps. Each showed a photograph of a particular formation, with brief details about the cloud type printed on the reverse of the backing sheet. The stamps were developed with help from The Weather Channel, the National Oceanic and Atmospheric Administration's National Weather Service, and the American Meteorological Society. But they came about thanks to the campaigning of Jack Borden (Member 009), from Athol, Massachusetts. "It took a decade and a half of all but outright begging," he told us, "before the USPS finally phoned to inform me that a block of cloud stamps would be issued. Persistence, plus letters I solicited from notables such as Senator Ted Kennedy, eventually prevailed." The goal of the souvenir stamps was to educate stamp collectors, young and old, about atmospheric sciences.

ABOVE "Cloudscapes" stamps issued in 2004 by the US Postal Service, thanks to the efforts of Jack Borden (Member 009).

WHENEVER YOU SEE an unfeasibly crisp and well-defined circle cut out of the sky, you can be pretty sure that you are looking up at a cavum cloud, sometimes known as a fallstreak hole. This feature also has a third name, a hole-punch cloud, and that is just how it looks: as if someone has punched a hole from the cloud with a celestial cookie cutter. The regularity of the curve, whether it be a complete circle or more of a cigar shape, gives the cloud an arresting appearance. This is due to the particular way it forms. Supercooled droplets within the cloud start to freeze in one region—set off perhaps by ice crystals falling into the layer from above, or by an aircraft climbing or descending through the layer. The resulting ice crystals grow rapidly and splinter as they do so to produce tiny seeds of ice. These then encourage the neighboring droplets to freeze too. It is a sort of chain reaction of freezing. The ice crystals fall below—appearing here as a graceful streak of white—and so leave behind a hole. Its sharp, geometric edge marks the distance to which the chain reaction of freezing has spread. Who gets to eat the enormous cloud cookie, nobody knows.

OPPOSITE Cavum cloud photographed by Pete Herbert (Member 32,675) over Lamington National Park, Queensland, Australia.

ISAAC LEVITAN WAS A CELEBRATED 19th-century Russian landscape painter who was relatively unknown in the West. He was a member of the independent group of artists known as *Peredvizhniki*, or "The Wanderers," for whom landscape paintings were a proud expression of Russian identity. Isaac Levitan became known for his "mood landscapes," which were brimming with feeling and atmosphere. "Can there be anything more tragic," the painter wrote in 1887 to his friend the playwright Anton Chekhov, "than the feeling of endless beauty in everything around you, to observe the hidden mysteries, to see God in everything and not to be able, realizing your inadequacy, to express all these great emotions adequately and fully?"

ABOVE Detail from *Before the Storm* (1890), by Isaac Levitan. Spotted by Andrew Pothecary (Member 3,769).

A CORONA IS CAUSED by the diffraction of light by tiny airborne particles such as cloud droplets, ice crystals, or even pollen. Since light often behaves as a wave, it bends slightly as it skims the edge of an obstacle like a cloud droplet. The amount by which it diffracts like this depends both on the light's wavelength and the size of the particle. Since different wavelengths bend by different amounts, the effect can result in a separating out of the spectrum to appear as rings of color around the Sun or Moon. The central blue-white disk of the corona is called the aureole. This part is common, and often has a reddish brown outer edge. The colored rings of a corona around the outside are a less common sight. The size of the corona as a whole depends on the size of the cloud particles. You could call it physics or you could call it a marvel. We like to think it is both.

ABOVE Corona produced by Cirrostratus and Cirrocumulus clouds, spotted over Álamos, Sonora, Mexico, by Suzanne Winckler (Member 41,844).

As the skies appear to a man, so is his mind. Some see only clouds there; some, prodigies and portents; some rarely look up at all; their heads, like the brutes', are directed toward Earth. Some behold there serenity, purity, beauty ineffable. The world run to see the panorama, when there is a panorama in the sky which few go to see.

From *The Journal of Henry David Thoreau*, January 17, 1852.

ABOVE Towering Cumulus and Cumulonimbus gilded by the setting Sun, spotted over Singer Island, Florida, US, by Luda Sinclair (Member 46,659).

OPPOSITE Astronauts aboard Space Shuttle Columbia in 1999 submit their winning entry for the "Most Clouds Spotted in One Go" competition.

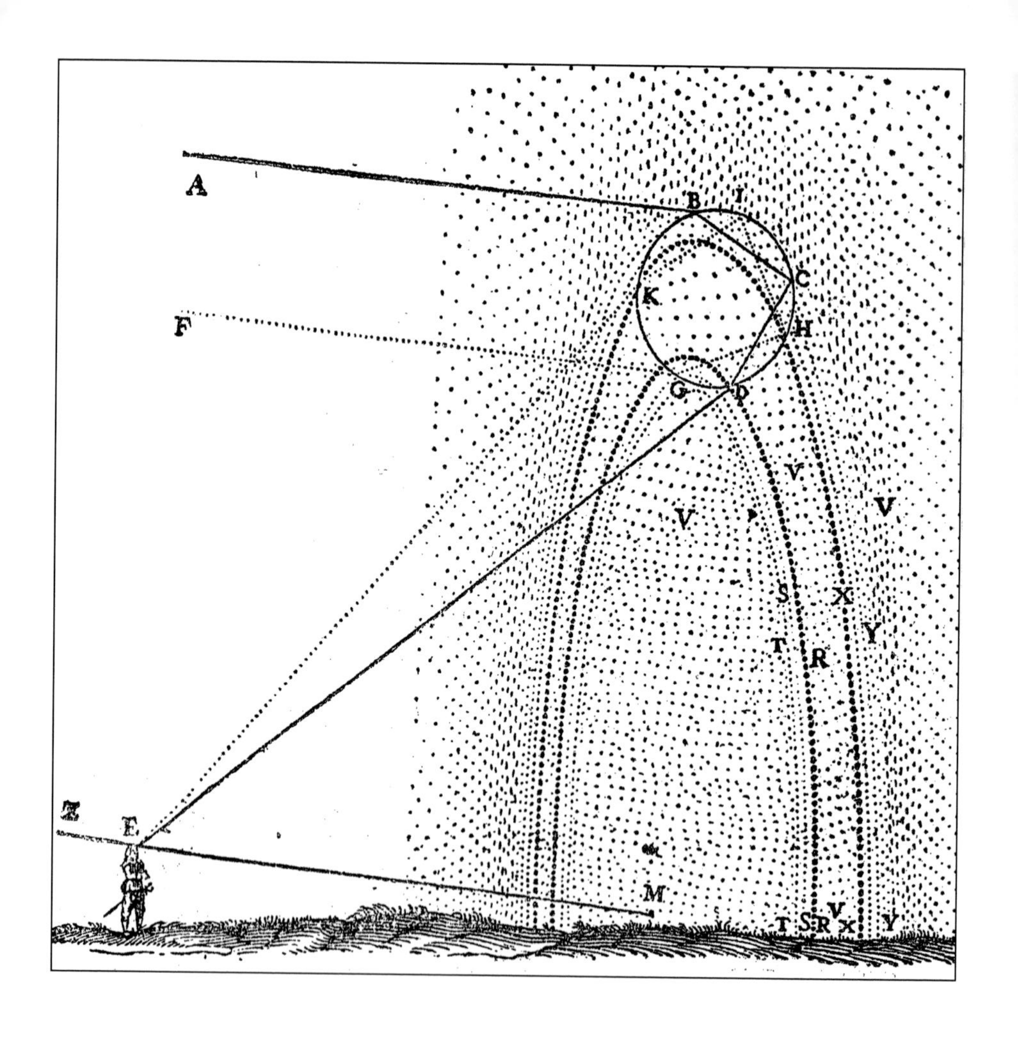

FRENCH NATURAL PHILOSOPHER René Descartes explained rainbows in his 1637 book *Discours de la Méthode*. He wasn't the first to realize they form by light passing through raindrops, but his experiments in shining light through a large glass globe of water to calculate the angles of refraction and reflection refined our understanding of the optics. In this illustration from the book the circle represents a raindrop and the straight lines rays of sunlight that bend, or refract, as they enter and leave the drops. It shows how light can form a primary bow or a secondary bow depending on whether it reflects once or twice within the drops. "If I can explain their nature," Descartes wrote about clouds in the book's appendix, " . . . one will easily believe that it is possible in some manner to find the causes of everything wonderful about Earth."

ABOVE Diagram of rainbow optics from *Discours de la Méthode* (1637), by René Descartes.

REMEMBER that the sky's an ever-present resource to turn to when you have too much on your mind.

ABOVE Cumulus, spotted by Tania Ritchie (Member 23,514) over Newcastle, New South Wales, Australia.

IN THE OPEN EXPANSE of the heavens there are few places for a cloud to hide. But one cloud type manages to remain largely overlooked simply by camouflaging itself against its much larger neighbors. Pannus is a cloud with the appearance of dark, ragged shreds. It is found clinging to the undersides of the rain-bearing beasts, Nimbostratus and Cumulonimbus. The cloud forms when the moisture-laden atmosphere under a precipitating cloud becomes so saturated that it condenses into Stratus-like patches. These look darker than the main cloud above simply because they block more of the already scant sunlight. When you spot the ragged forms of pannus beneath another cloud the chances are it is either raining, has just rained, or the rain will begin within minutes.

ABOVE Pannus beneath a rain cloud, spotted over Hatfield, Hertfordshire, England, by Justin Parsons (Member 15,125).

TOP In August 2017, millions in the US watched the total solar eclipse from below. Only six people watched it from above. This photograph by one of the six astronauts aboard the International Space Station at the time shows the Moon's shadow, known as its umbra, as it sweeps across the American skies.

BOTTOM The Southern Lights, or aurora australis, are just as spectacular as the northern ones. They receive less attention because they form mostly out over open ocean. When occasionally viewed from land, like these spotted from Victoria, Australia, they appear off on the southern horizon.

THE CLASSIC 1726 NOVEL *Gulliver's Travels* by Jonathan Swift describes a great floating island named Laputa, which can be moved about by its inhabitants using a form of magnetic levitation. No doubt Swift got the idea from a patch of solid-looking Stratocumulus like this one. Swift describes the Laputans as being obsessed with mathematics, astronomy, and music. In fact, so lost are they in thought that they cannot function in everyday life unless constantly struck by a bladder full of pebbles to wake them from their daydreams. Each Laputan is therefore escorted by a servant, called a clapper, who performs this essential role. Sounds like a useful service.

ABOVE Stratocumulus, spotted by Charlie Gray (Member 7,670) over Carvoeiro, Faro, Portugal.

CLOUD SHADOWS RENDERED VISIBLE by haze in the atmosphere are known as crepuscular rays. They appear to burst out from a cloud like this towering Cumulus when you are observing them as the sunlight shines towards you. The layer of atmospheric haze onto which the cloud casts its shadow is in fact below the cloud's summit. The shadow is therefore nearer to you. The effect of perspective means that the nearer shadow looks larger than the cloud and the rays appear to fan outwards as they approach.

ABOVE Crepuscular rays appearing to burst from a Cumulus congestus cloud, spotted by Tiziano Bartolucci over Rieti, Lazio, Italy.

THE WORLD'S BEST LIGHTNING DISPLAYS occur in Venezuela where the Catatumbo River empties into Lake Maracaibo. These pyrotechnic spectacles are a result of local topography, wind patterns, and the location within the globe's often stormy Intertropical Convergence Zone. The outstanding feature of the abundant storm clouds over this region is that they occur in the same place and at the same time for nearly half of the nights throughout the year. The thunderstorms are so high and far enough away from local settlements that they are often seen without any sound of thunder. The silent storms produce so much lightning for up to ten hours a night that the locals need blackout blinds to sleep.

ABOVE Cumulonimbus with lightning, spotted by Fernando Flores over the Catatumbo River, Venezuela.

THE BLUR BUILDING, created by architects Diller Scofidio + Renfro for the 2002 Swiss Expo, had walls made of fog. This temporary structure consisted of a lightweight frame covered with 31,500 nozzles, through which a fine mist of water was pumped from Lake Neuchâtel below. Visitors entered via a walkway across the lake. The water pressure of the jets was computer-controlled, taking into account the temperature, humidity, and wind conditions, to ensure that for the five months of its existence the Blur Building was forever shrouded in a swirling, shifting cloud.

ABOVE Architects Diller Scofidio + Renfro's Blur Building, made from a cloud for the Swiss Expo of 2002.

THE 17TH-CENTURY DUTCH PAINTER Aelbert Cuyp never traveled far from his native Dordrecht. In his 1655–1660 painting *Landscape with a View of the Valkhof, Nijmegen*, the artist captured an unusual formation of Stratocumulus castellanus exhibiting the distinctive cloud holes known as lacunosus. Nice. Skyscapes more than landscapes remind us that you don't need to rush across the world to be surprised. Just step outside and pay attention to the everyday stuff that most people miss.

ABOVE Stratocumulus castellanus lacunosus, spotted by Aelbert Cuyp in this detail from his *Landscape with a View of the Valkhof, Nijmegen*.

OPPOSITE Cirrus uncinus, spotted over New Plymouth, Taranaki, New Zealand, by Graham Billinghurst (Member 24,513).

THE CLOUD SPECIES HERE IS UNCINUS, a form of Cirrus whose name is from the Latin for "hooked." This is when the high cloud's long streaks of ice crystals hook up at one end. Counterintuitively, the wind causing the formation is blowing from left to right, not the other way around. At the hooked end, where the ice crystals start to fall, the wind is blowing very fast to the right. As they descend, they pass into winds still blowing to the right but at a much slower rate. The sudden drop in speed with altitude, known as wind shear, means the crystals trail further and further behind with their descent. Dramatic wind shear like this one, revealed by the "mares' tail" appearance of Cirrus uncinus, indicates that a weather front is on its way.

O! it is pleasant, with a heart at ease
Just after sunset, or by moonlight skies,
To make the shifting clouds be what you please . . .

From "Fancy in Nubibus or the Poet in the Clouds" (1819), by Samuel Taylor Coleridge.

ABOVE A piglet surfing on the River Great Ouse at King's Lynn, England, spotted by Matt Minshall (Member 7,721).

ABOVE A tornado extends from the highly turbulent inflow region of a supercell storm system. Generally situated to the rear of the advancing storm, this inflow region is often marked by a wall cloud, or murus, which fills the upper part of this image. Spotted over Katie, Oklahoma, US, by Dave Hall (Member 840), this is what it looks like when a storm cloud wants to feel the ground.

THE THEATRE OF EPIDAURUS is the sort of place where a classic Greek comedy like *The Birds* might have been performed. Written in 414 BC by the playwright Aristophanes, it is about two gentlemen who are fed up with the hurly-burly of Athens. Desperate for a little peace and tranquility, they persuade the birds to build them a utopian city to escape to up in the clouds. This turns out, of course, to be not the most practical of plans, and things generally end badly for them. But the two characters do come up with a good name for their bird-made city in the sky: *Nephelokokkygia*. The Ancient Greek is translated as "Cloud Cuckoo Land," a place associated ever since with those whose dreams float free.

ABOVE Stratus takes to the stage at the Theatre of Epidaurus, near Lygourio, Greece, spotted by Martin Foster.

VAN GOGH PAINTED HIS MOST interesting skies in the last three years of his life, after he moved to Provence in the south of France in 1888. This painting, for instance, shows a lenticularis cloud over the Alpilles mountains, with the distinctive trail of ice crystals that this cloud formation can sometimes exhibit. Back in the vicinity of Paris in 1890, Van Gogh wrote to his brother Theo, "I almost think these canvases will tell you what I cannot say in words, the health and restorative forces that I see in the country." Two weeks later, he took his life, on a day of fine weather at the end of July.

ABOVE Detail from *The Olive Trees* (1889), by Vincent van Gogh.

(12)

The Clouds. VIII. *Nubes.*

A Vapour, 1. *ascendeth from the* Water.	Ex *Aqua* ascendit *Vapor*, 1.
From it a Cloud, 2. *is made, and a* white Mist, 3. *near the Earth.*	Inde fit *Nubes*, 2. & propè terram *Nebula*, 3.
Rain, 4. *and a* small Shower *distilleth out of a Cloud, drop by drop.*	E *Nube* guttatim stillat *Pluvia* 4. & *Imber*.
Which being frozen, is Hail, 5. *half frozen is* Snow, 6. *being warm is* Mel-dew.	Quæ gelata, *Grando*, 5. semigelata, *Nix*, 6. calefacta, *Rubigo* est.
In a rainy Cloud, set over against the Sun, the Rainbow, 7. *appeareth.*	In nube pluviosâ, Soli oppositâ, apparet *Iris*, 7.
A drop *falling into the water, maketh a* Bubble, 8. *many* Bubbles *make froth*, 9.	*Gutta* incidens in aquam facit *Bullam*, 8. multæ *Bullæ* faciunt spumam, 9.
Frozen Water is called Ice, 10. Dew congealed,	Aqua congelata *Glacies*, 10. *Ros* congelatus,

THE *ORBIS PICTUS* WAS ONE of the first printed picture books for children. Published in 1658, it was written by the Czech educationalist John Amos Comenius. The book explains a wide range of subjects, both natural and man-made. Here are the annotations to its woodcut illustration about clouds:

A Vapour, 1. ascendeth from the Water. From it a Cloud, 2. is made, and a white Mist, 3. near the Earth. Rain, 4. and a small Shower distilleth out of a Cloud, drop by drop. Which being frozen, is Hail, 5. half frozen is Snow, 6. being warm is Mel-dew. In a rainy Cloud, set over against the Sun, the Rainbow, 7. appeareth. A Drop falling into the water, maketh a Bubble, 8. many Bubbles make froth, 9. Frozen water is called Ice, 10. Dew congealed is called a white Frost. Thunder is made of a brimstone-like vapor, which breaking out of a Cloud with Lightning, 11. thundereth and striketh with lightning.

ABOVE "Clouds" from *Orbis Pictus* (1658), by John Amos Comenius, spotted by Heather Silverwood (Member 42,036).

OPPOSITE Altocumulus lenticularis, spotted over Lake Pukaki, South Island, New Zealand, by Tania Ritchie (Member 23,514).

LENTICULARIS ARE OROGRAPHIC CLOUDS, which means they result from the interaction between wind and raised terrain like hills or mountains. This display was caused by the Southern Alps of New Zealand's South Island. When atmospheric conditions are stable, the airflows downwind of peaks can take on rising and dipping paths. They flow in invisible standing waves of air like those of water on the surface of a fast-flowing river just downstream of a rock. Whenever air rises it expands, and gases that expand cool, so the air cools at the crests of these wind waves. Such movements are invisible unless air conditions are just right for the cooling to encourage the air's moisture to condense into droplets. These we see as smooth, saucer- and lozenge-shaped lenticularis clouds hovering in place in the wind.

ABOVE The word "cloud" stems from the Old English word *clūd*, meaning "boulder." No wonder this Cumulus looks so at home with the 6,000-year-old megaliths at Carnac, Brittany, France. Spotted by Harriet Aston (Member 42,078).

OPPOSITE Cirrus uncinus, spotted over Place de la République, Paris, France, by Marjorie Perrissin-Fabert (Member 32,721).

Tell me, enigmatic man, whom do you love best? Your father, your mother, your sister, or your brother?

"I have neither father, nor mother, nor sister, nor brother."

Your friends, then?

"You use a word that until now has had no meaning for me."

Your country?

"I am ignorant of the latitude in which it is situated."

Then Beauty?

"Her I would love willingly, goddess and immortal."

Gold?

"I hate it as you hate your God."

What, then, extraordinary stranger, do you love?

"I love the clouds—the clouds that pass—yonder—the marvellous clouds."

From *The Stranger* (1862), by Charles Baudelaire.

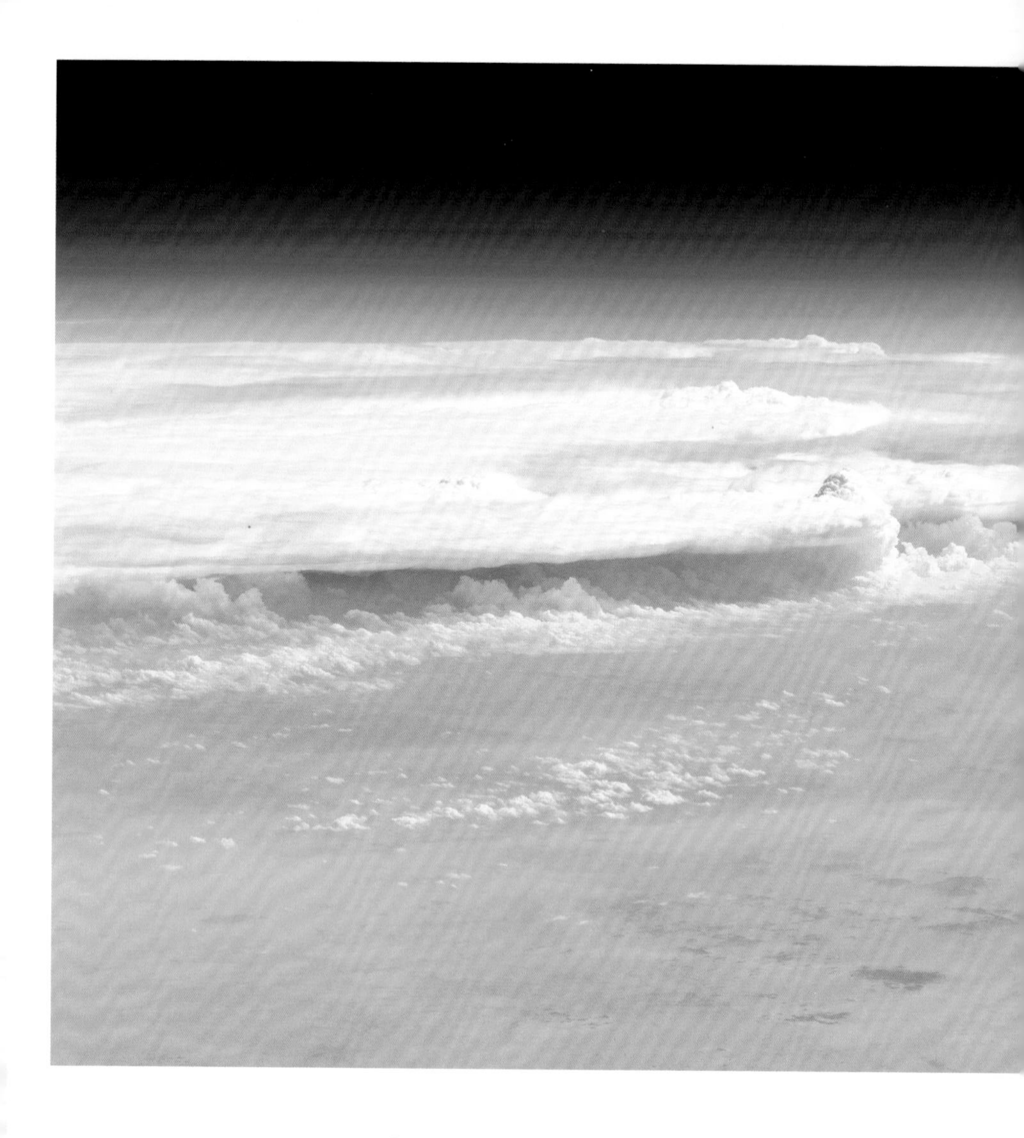

THIS IS THE VIEW OF A STORM front from above, and it shows just how far the canopy at the top of a Cumulonimbus can stretch out. It might extend as far as 160 kilometers (100 miles) from the storm center. This upper part of the cloud is often described as resembling an anvil, which is what its Latin classification *incus* actually means. In this case, it is a very long anvil—no doubt, one on which shoes are forged with a thunder hammer for the mares whose tails dance across far fairer skies.

ABOVE A white buffalo sleeps over eastern Colorado, US, spotted by Patrick Dennis (Member 43,666). The buffalo is lying on a flat, invisible floor of air that marks the condensation level. This is the height at which air, rising off the Sun-warmed ground as a thermal, cools enough for its moisture to change from gas to liquid droplets. A Cumulus cloud results, the thermal made visible. A snoozing beast of moisture.

OPPOSITE Cumulonimbus clouds along a storm front near the city of Medina in the Al-Qassim region of Saudi Arabia, spotted by astronauts aboard the International Space Station.

CLOUDS HELP MODERATE global temperatures—but in a complex way. Low clouds have an overall cooling effect compared to clear skies. They reflect away more of the Sun's heat than they trap in of the Earth's. High clouds have an overall warming effect. They trap in more of the Earth's heat than they reflect away of the Sun's. Averaging it all out, Earth's clouds keep surface temperatures cooler than they'd be if we had none. But we have little appreciation of how clouds might change with rising global temperatures, so who knows if things will average out the same way in the future?

TOP A crescent Earth, spotted from NASA's unmanned Apollo 4 test flight in 1967, at an altitude of almost 16,000 kilometers (10,000 miles).

RIGHT A bear marvels at the spectacle of Cirrocumulus above. Spotted by Judy Taylor (Member 42,887) over Kathmandu, Nepal.

OPPOSITE An ocean of air spotted over the Italian Alps from the flight deck by Richard Ghorbal (Member 5,117).

Let us suppose for a moment that a being, whose eyes were so made that he could see gases as we see liquids, was looking down from a distance upon our earth. He would see an ocean of air, or aerial ocean, all round the globe, with birds floating about in it, and people walking along the bottom, just as we see fish gliding along the bottom of a river . . .

From "The Aerial Ocean in Which We Live," *The Fairy-Land of Science* (1883), by Arabella B. Buckley.

ABOVE Swirling ammonia clouds of planet Jupiter, spotted by NASA's Juno spacecraft while orbiting at a distance of 18,906 kilometers (11,747 miles). They are extraterrestrial abstract art.

OPPOSITE Sun pillar, spotted in the evening by Thorleif Rødland over Fensfjorden, Norway.

A SUN PILLAR is like the airborne equivalent of the glitter path that appears when the Sun reflects off ocean waves. It is a vertical column of light stretching above or below the Sun that is caused by the collective glints from the surfaces of countless tiny ice crystals suspended up in the clouds. Since the sunlight need not pass through the crystals to form a sun pillar, this optical effect is far less fussy than the other halo phenomena. It doesn't depend on particular shapes and exact orientations of cloud ice crystals, nor ones that are clear and faultless enough for the light to shine right through. So long as the Sun is low in the sky or just below the horizon, any ragtag of flattish ice crystals falling like autumn leaves will do a great job of building a rosy pillar as a monument to the passing day.

ABOVE Stratocumulus and Cumulonimbus clouds show their silver linings when lit from behind. Spotted over central Spain, the *View of Toledo* (c. 1599) was painted by Doménikos Theotokópoulos, who, on account of his Greek origins, is more commonly known as El Greco.

OPPOSITE Crepuscular rays shining through gaps in Stratocumulus clouds, spotted by Jelte van Oostveen (Member 38,512) over Futrikelv, near Tromsø, Norway.

The sun, moving as it does, sets up processes of change and becoming and decay, and by its agency the finest and sweetest water is every day carried up and is dissolved into vapor and rises to the upper region, where it is condensed again by the cold and so returns to the earth. This, as we have said before, is the regular course of nature.

From *Meteorologica* by the Ancient Greek philosopher Aristotle, 350 BC.

THIS 19TH-CENTURY PHOTOGRAPHIC PRINT of Cumulus clouds over the Mediterranean Sea at Sète in southern France may not look particularly fancy, but it was in fact groundbreaking. It was produced by French artist and pioneer of early photography Gustave Le Gray. Up until the 1850s, achieving a photographic exposure that would work both for the sky and the landscape (or seascape) had proved impossible due to the limitations of camera technology. Le Gray solved this challenge by printing from two separate negatives. One was exposed to the sky and the other to the sea. This allowed him to achieve a balance of tones across the whole scene. Le Gray's prints produced in this manner were acclaimed internationally for their impressive depth of light and shadow.

ABOVE Detail of *The Great Wave, Sète,* by Gustave Le Gray.

THIS MULTICOLORED CIRRUS CLOUD appeared to David Rosen and family at noon over Richardson Bay. It joined them for almost an hour. It was effectively a tiny peek at the large optical effect known as a circumhorizon arc. Caused by the light from a high Sun shining through prism-shaped ice crystals, this halo phenomenon often appears as a huge, flat band of colors stretching across the sky near the horizon. But that is only when the right kind of ice crystals are spread over a wide enough area of the sky. When they're concentrated into a single cascade of Cirrus like this, just a small view of the optical effect appears where the ice crystals are falling—a glimpse of the broad light show as if through a celestial keyhole.

ABOVE Circumhorizon arc, spotted by David Rosen (member 23,717) over Richardson Bay, California, US.

THERE'S A SKY VERSION of those contour lines you see on maps to delineate mountain terrain. You can see it in certain forms of lenticularis cloud, particularly when they are lit at a glancing angle like these ones. While the classic lenticularis cloud appears as a solitary disc, the formation does sometimes have a more continuous, meandering shape. The edges delineate where undulating airflows downwind of mountains rise and cool enough for cloud to form. The contour-line effect at the edges of the lenticularis happens when the flowing air is stratified, with layers of moister air sandwiched between those of drier air, because cloud forms more readily in the moister layers than the drier. These clouds demonstrate how, in stable conditions, our lower atmosphere flows in sympathy with the physical contours below.

Let us a little permit Nature to take her own way; she better understands her own affairs than we.

From "Of Experience," *Essays* (1580), by Michel de Montaigne.

ABOVE Cavum, also known as fallstreak holes, spotted over Half Moon Bay, California, US, by Paul Jones (Member 18,562).

OPPOSITE Altocumulus lenticularis at sunset, spotted by Gary Davis (Member 21,168) over Varese in the Italian Alps.

ACCORDING TO SOME ACCOUNTS, Ancient Hindu creation myths held that the world began with the help of mythical elephants who were white, could fly, and had the power to bring rain. Prayers to albino elephants asking for help during times of drought would sometimes refer to the animals by the name of *Megha*, which means "cloud." Having conducted an extensive analysis of the many thousands of clouds in the shape of things on the Cloud Appreciation Society photo gallery, we can confirm that the animals spotted most frequently by our members are elephants. What does all this mean? We have absolutely no idea.

OPPOSITE PAGE Elephants in the sky, spotted by (**CLOCKWISE FROM TOP LEFT**): Fred Ohlerking (Member 41,191) over Yosemite National Park, California, US; Graham Blackett (Member 928) over Phuket, Thailand; Anne Downie (Member 12,153) over Annan, Dumfriesshire, Scotland; Lauren Antanaitis (Member 25,124) over Cornelius, North Carolina, US.

THIS PAGE (**CLOCKWISE FROM TOP LEFT**): Hélène Condie (Member 28,830) over Ottawa, Ontario, Canada; Saskia van der Sluis (Member 23,801) over Ameland, Waddenzee, Holland; Peter Beuret (Member 36,471) over Myrtle Beach, North Carolina, US; Sugata Kuila over Havelock Island, Andaman, India.

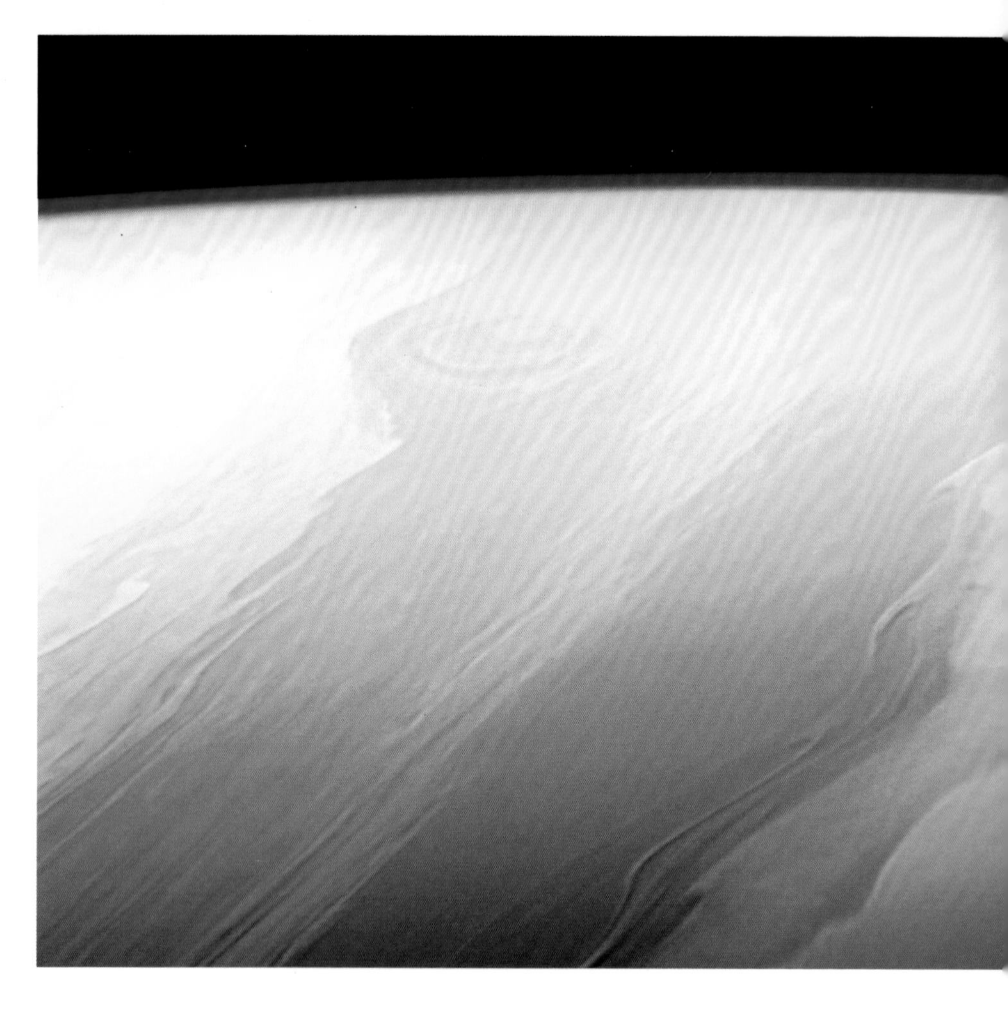

THE CLOUDS IN SATURN'S ATMOSPHERE vary greatly with altitude. The lowest cloud layer is made up of water ice and extends some 10 kilometers (6 miles) in height. A cloud layer made up of ammonium hydrosulfide ice extends 50 kilometers (30 miles) above this. A further 80 kilometers (50 miles) above that, and the clouds are made of ammonia ice. It is the tops of these that we can see through the hydrogen and helium haze of the upper atmosphere (shown here in blue). Neighboring bands of cloud move at different speeds and directions depending on their latitudes, which generates turbulence where they meet and leads to the wavy structure along the interfaces. By studying Saturn's clouds scientists can improve their theories of fluid motion and better understand the atmospheres of other planets—including our own.

. . . the sky, with silver swirls like locks of toss'd hair, spreading, expanding—a vast voiceless, formless simulacrum—yet may-be the most real reality and formulator of everything—who knows?

From "A July Afternoon by the Pond," *Specimen Days* (1892), by Walt Whitman.

ABOVE Cirrus uncinus, spotted by Carole Pereira over Rossmoyne, near Perth, Western Australia.

OPPOSITE The clouds of Saturn, spotted from 1.2 million kilometers (750,000 miles) away by NASA's Cassini spacecraft.

THE 18TH-CENTURY FLEMISH ARTIST Simon Denis was an early practitioner of outdoor landscape painting. These towering Cumulonimbus storm clouds over Rome are one of 48 cloud studies he painted in the late 18th century to hone his sky skills. A hundred years later, such free and expressive skyscapes would have been exhibited as finished paintings in their own right. For Denis, a sketch made outdoors like this was no more than a reference, something to be hidden away in his studio and later incorporated into the backgrounds and margins of exhibited works. For us cloudspotters, these sketches are the real deal—where the dynamic power of nature takes center stage.

ABOVE *Study of Clouds with a Sunset near Rome* (1786–1801), by Simon Denis, spotted by Karen Shuker (Member 45,918).

THE ICE CRYSTALS in feather-shaped Cirrus refract the sunlight to produce a circumzenithal arc, the high optical effect whose colors are purer than those of a rainbow. Either that, or a bird of paradise from the upper troposphere has shed its tail feathers to the wind.

ABOVE Circumzenithal arc formed by Cirrus vertebratus, spotted by Beth Holt over Yuma, Arizona, US.

THIS BRIGHT TRAIL spotted by Colleen Thomas is a rare and unusual example of a noctilucent cloud. At altitudes of 85 kilometers (53 miles), these night-shining clouds form much higher than normal weather clouds, from where they can still catch the sunlight when the lower sky is dark. Noctilucents are unusual, mysterious formations, and the one spotted here over California was particularly so. This night-shining cloud didn't have the formation's usual bluish, ghostly, rippled appearance. It looked more like a twisted aircraft condensation trail. It was also in the wrong part of the globe. Noctilucent clouds usually form much nearer the Poles. San Francisco, at a latitude of just over 37 degrees, is way outside the latitudes of 50–70 degrees where they are normally observed. It even appeared at the wrong time of year. Night-shining clouds are generally spotted in the summer months, which is when the upper atmosphere, counterintuitively, is at its coldest. Why then would one appear so far south, and in the middle of winter? This particular night-shining cloud was in fact a condensation trail—formed not by an aircraft, or even a rocket, but by a meteor. As the meteor burned up in the high atmosphere, it produced specks of meteor dust, and these acted as tiny seeds onto which the ice crystals of the cloud could start to freeze. The meteor likely disintegrated long before reaching the ground, but in its wake hung the rarest of sights for Californian skies: a noctilucent trail shining bright in the twilight; ice, born of a meteor, adrift in the currents of the upper mesosphere.

OPPOSITE Noctilucent cloud, spotted by Colleen Thomas over San Francisco, California, US.

THE GREENS AND PINKS of the aurora appear to fan out across the sky like this when they form directly overhead. The particular pattern in the mesmerizing, rippling clouds of light emission occurs in regions of the world known as the aurora ovals that ring the north and south magnetic poles. The ovals are where the Earth's magnetic field lines pass directly down through the surface towards its magnetic core. The streaks of light follow the field lines and, like so many dramatic appearances of the sky, radiate out like this as they approach due to the effect of perspective.

There are no rules of architecture for a castle in the clouds.

From *The Everlasting Man* (1925), by G. K. Chesterton.

ABOVE Pileus, Velum, Cumulus, and Altocumulus float around a mighty structure of a Cumulonimbus storm cloud. Spotted over Choctawhatchee Bay, Florida, US, by Vicki Kendrick (Member 39,727).

OPPOSITE Aurora borealis, spotted by Juho Holmi over Første Fiskfjordvatnet, Hinnøya, Norway.

YOU CAN THINK OF FOG as the lowest of all cloud forms. It is when the featureless layer cloud known as Stratus forms right down at ground level. Purists would argue that altitude is an essential part of being a cloud, and that fog therefore can't be a bona fide cloud. At the Cloud Appreciation Society, we are not purists. We appreciate fog as the only cloud prepared to come and visit us on the surface, and because it reveals the beauty of a landscape by hiding it.

ABOVE December fog, spotted from the window seat by Easton Vance (Member 3,464) over Puerto del Rosario, Canary Islands, Spain.

OPPOSITE Detail from *The British Channel Seen from the Dorsetshire Cliffs* (1871), by John Brett.

WHEN THE BRITISH ARTIST John Brett painted a view of the sea from the south coast of England in 1871, he showed fair-weather Cumulus clouds and crepuscular rays of sunlight shining down onto the water's surface. Although Brett clearly loved the play of light in the atmosphere, he was no cloudspotter. The sky above doesn't match the light on the sea below. The artist must have combined the sky from one day with the sea from another. Such patches of light on the water's surface would appear when a layer of low cloud called Stratocumulus covers most of the sky, but for gaps here and there that let the Sun shine through. They would never result from the faint shadows cast by a gentle scattering of Cumulus like this. Nice try, Brett, but no cloudspotting cigar.

THE IC 2118 NEBULA is about 800 or 900 light years from us. This gas cloud is situated near the constellation of Orion, just next to the hunter's knee as it happens. Dust in the cloud reflects light towards us from the bright star Rigel. If you think IC 2118 is a less-than-catchy name, you wouldn't be alone. This interstellar cloud is now known more commonly as the Witch Head Nebula. Take another look, and you'll see why.

ABOVE The IC 2118 nebula, spotted by NASA as part of the STScI digitized Sky Survey.

OPPOSITE Volutus, also known as a roll cloud, spotted by Ross Hofmeyr as it arrives ashore at Overstrand, Western Cape, South Africa.

Go forth under the open sky, and list to Nature's teachings.

From "Thanatopsis" (1817), by William Cullen Bryant

THE EXTREME TEMPERATURES, mountain terrain, and high winds of Antarctica ensure it has more than its fair share of nacreous clouds. Also known as polar stratospheric clouds, these can exhibit dramatic bands of iridescent colors. They were admired by the explorers of Captain Scott's ill-fated 1911 expedition to the South Pole. The expedition doctor, Edward Adrian Wilson, painted these from the base camp on Cape Evans before they set off for the Pole. Wilson, along with the other four team members, would never return.

ABOVE Watercolor of nacreous clouds over Cape Evans, Antarctica, painted in 1911 by Dr. E. A. Wilson, the doctor on the Terra Nova expedition to the South Pole.

OPPOSITE Fluctus over the Rocky Mountains, spotted by Hallie Rugheimer (Member 35,218) from Shields Valley, Montana, US.

THE ROCKY MOUNTAINS OF MONTANA, US, are transformed into giant waves as the low Stratus cloud develops wave features known as fluctus. These breaking-wave shapes develop in conditions of shearing winds, when the wind speed increases markedly with altitude along the upper boundary of the cloud layer. The shearing effect causes the upper edge of the cloud to become ruffled into rising and dipping undulations. When the wind speeds are just right, the tops of the undulations can be curled over into vortices. Mountain ranges are great for setting up shearing wind patterns because the airflow at lower altitudes is slowed as it encounters the mountain range while the air higher up flows unimpeded.

Study nature, love nature, stay close to nature.

It will never fail you.

Architect Frank Lloyd Wright's advice to his students.

ABOVE Fog at sunrise, spotted over Pleasant Valley, Vermont, US, by Kristina Machanic (Member 38,409).

ABOVE A "rain deer," spotted over Bampton, Oxfordshire, England, by Marie Dent (Member 9,934).

CIRRUS-LIKE ICE CLOUDS at sunrise over Valles Marineris. The scene is around 1,000 kilometers (600 miles) across. Although the clouds are over the equator, they are a long, long way from anywhere on Earth, as they are water-ice clouds over equatorial Mars.

SINCE CIRRUS CLOUDS are named after the Latin word for a lock of hair, the Cirrus fibratus variety are presumably when the locks have just been neatly combed before school.

ABOVE RIGHT Ice clouds, spotted over Valles Marineris, Mars, in 1976 by NASA's Viking 1 Orbiter.

RIGHT Cirrus fibratus, spotted by Althea Pearson (Member 38,865) over Somerton, Somerset, England.

OPPOSITE Detail from *Bleaching Ground in the Countryside Near Haarlem* (1670), by Jacob van Ruisdael, Kunsthaus Zürich art museum, Switzerland, spotted by Hans Luchies (Member 41,544).

THEY SAY THAT CLOUDS are the Dutch mountains, and never has this been better expressed than in the 17th-century landscape paintings of Jacob Isaackszoon van Ruisdael. He is considered the preeminent landscape painter of the Dutch Golden Age, and he liked to set out his canvases with low horizon lines so that he could fill them with clouds. His impressively realistic skies had a major influence on the development of Western landscape art. Perhaps the artist picked up his love of clouds from his father, Isaack van Ruisdael, a landscape painter too, who was partial to a dramatic sky. Or maybe it was from his landscape-painting uncle, Salomon van Ruysdael. Uncle Salomon was more famous than his brother—no doubt, because he took more care over his clouds, which were almost as voluminous as those of Nephew Jacob.

ABOVE Clouds can cast very long shadows. Here a distant Cumulonimbus, hidden on the horizon, makes its presence known for just a few moments at sunset as it casts a shadow right across the Altocumulus sky. Spotted by Baiyan Huang over Zhejiang, China.

OPPOSITE A tuba descending from a storm system over Keenesburg, Colorado, US, spotted by Carlye Calvin (Member 45,668).

A FINGER OF CLOUD descending within a vortex of air from the base of a Cumulonimbus is known as a tuba. Like water rotating down the plughole of a bath, the air can develop this spin as it rushes up into the base of the building storm. The air pressure drops at the center of the vortex. This causes the air temperature to drop too, and that can be enough to encourage the humid air to form droplets. As the humid air is sucked up with growing ferocity, feeding into the belly of the mighty Cumulonimbus, it spins faster and faster and the umbilical form of the tuba extends downwards. We all know what can happen when it touches down.

E 14

THE *NIMBUS* SERIES of works by contemporary artist Berndnaut Smilde are clouds created indoors. Each lasts for just a few seconds. The Nimbus is captured on camera before the moment is gone. Smilde creates the cloud by saturating the air within the space with a fine mist of water and then introducing a puff of smoke. The water condenses onto the smoke particles just like the droplets of a natural cloud form around tiny condensation nuclei in the atmosphere, such as particles of dust, ash, and organic compounds. Each *Nimbus* photograph functions as a document of something that happened in a specific location for a fleeting moment and is now gone. "I see them as temporary sculptures," says the artist, "made of almost nothing, balancing on the edge of materiality."

LEFT *Nimbus Dumont,* 2014, by Dutch artist Berndnaut Smilde.

ACCORDING TO NORSE MYTHOLOGY, Frigg (sometimes known as Frigga) was goddess of the atmosphere. She was married to the mighty Odin, which made her queen of the gods. Frigg had her own palace called Fensalir, complete with a hall of mists and sea. She was able to predict the future but tended to keep it to herself. The word for "Friday" is named after her in Germanic languages such as English. Frigg had a fancy jewelled spinning wheel in her hall of mists, with which she used to spin long, bright threads of cloud. This is how Cirrus clouds used to be made—and, quite possibly, still are.

ABOVE A Cirrus skydiver, spotted by Lilian van Hove, aims for splashdown in a fjord near Tromsø, Norway.

OPPOSITE *Frigga Spinning the Clouds* by J. C. Dollman, from the book *Myths of the Norsemen* (1922), by H. A. Guerber.

NO ONE WOULD EVER SAY Nimbostratus is the most attractive of clouds. This dark, featureless wet blanket of the sky is surely the least popular of all the types. It is the one that gives other clouds a bad name. But the prolonged and steady precipitation from Nimbostratus gets the job done—converting salt water in oceans to fresh water on land, feeding plants, you know, sustaining life. Here's one doing just that.

ABOVE Nimbostratus, spotted by Hannah Hartke over Albemarle, Virginia, US.

OPPOSITE Primary, secondary, and reflection rainbows, spotted over Loch Dunvegan, Isle of Skye, Scotland, by Mike Cullen (Member 23,089).

AS WELL AS A STRONG PRIMARY RAINBOW and a fainter outer secondary bow, this image also shows a rare reflection bow, which appears between the two and bent at an odd angle. Like its regular rainbow companions, this elusive optical effect is formed by the reflection and refraction of sunlight shining from behind the viewer onto a shower of raindrops up ahead. But the reflection bow differs in that it is formed from sunlight that first reflected up off a body of water behind the viewer. The reflection means that it is as if this rainbow is being caused by a Sun below rather than above the horzion, which accounts for its odd angle. If you are ever lucky enough to see one, you'll find that a reflection bow always meets its regular-rainbow counterpart at exactly the level of the horizon.

There it is, fog, atmospheric moisture still uncertain in destination, not quite weather and not altogether mood, yet partaking of both.

From *Sundial of the Seasons* (1964) by Hal Borland.

ABOVE *Winter Landscape 2* (2007), by Alex Katz.

OPPOSITE The Helix Nebula, spotted with the 0.8-meter telescope at McDonald Observatory, Texas, US.

THIS IS AN INTERSTELLAR CLOUD. It is known as the Helix Nebula, and it consists of dust, hydrogen, helium, and other ionized gases. They are centered like a wreath around a dying star. As it runs out of nuclear fuel, the star has blown off its outer atmosphere. The blue-green light at the interior is emitted by oxygen, and the red light further out by hydrogen. Don't get worried, but one day in about five billion years, our Sun will meet the same fate.

THE ODD-SHAPED WEDGE of cloud sloping down from the underside of this storm system was spotted over Canberra, Australia. Known as a murus, or wall cloud, it can be thought of as the rear bumper of a supercell storm. The cloud feature develops behind where all the precipitation is falling, and it marks the inflow region of the storm system where warm moist air is being sucked up to feed the beast. A wall cloud is liable to sprout tubas, the rotating columns of cloud that can develop into land- and waterspouts, as well as fully fledged tornadoes. Tailgating this particular cloud feature is not to be recommended.

ABOVE Murus, or wall cloud, spotted by Wayde Margetts (Member 37,625) over Canberra, Australia.

Those who contemplate the beauty of the earth find reserves of strength that will endure as long as life lasts . . . There is something infinitely healing in the repeated refrains of nature—the assurance that dawn comes after night, and spring after winter.

From *The Sense of Wonder* (1965), by Rachel Carson.

ABOVE A sun pillar over Alpine, near Nuiqsut, North Slope Borough, Alaska, US, spotted by James Helmericks (Member 19,987).

ALTOCUMULUS CLOUDS OFTEN APPEAR in orderly patterns of clumps, like those shown opposite. The reason for this regularity is illustrated in the photograph below it. This shows a kitchen demonstration using the hotplate of a panini toaster onto which was poured a thin layer of vegetable oil mixed with some glitter powder. The powder reveals the oil's movement. The toaster was turned on very briefly to heat the oil gently from below. As it warms in contact with the hotplate, the oil floats upwards and the cooler oil above sinks back down to replace it. The interesting part is that the movement arranges itself into a regular pattern of cells of rising oil separated by gaps of sinking oil. This pattern just develops naturally, and it's what happens also in the air when regular patterns of Altocumulus clouds develop. Warmer air rises from below the cloud layer as cooler air sinks down to replace it from above. Just as with the oil, the air can't move up and down en masse; it needs to arrange itself into regions of rising and regions of sinking, which are more officially known as convection cells. Clumps of cloud appear in the rising cells. Gaps appear in the sinking regions between. Order emerges—if only for a few moments—in the glorious chaos of the sky (and the less glorious chaos of the kitchen).

OPPOSITE ABOVE Altocumulus, spotted by Brett King over Maroochydore, Sunshine Coast, Queensland, Australia.

OPPOSITE BELOW A kitchen demo shows the same pattern as oil is gently warmed in a panini toaster.

SUPERNUMERARY BOWS are repeating fringes of colors that can sometimes appear along the edge of a rainbow. They form at the inside of the primary rainbow. Where an outer, secondary bow is also present, faint supernumeraries can sometimes be seen along its outer edge. The fringes are caused by the interference of light waves emerging from the raindrops leading to brighter and darker bands. Supernumerary bows only appear bright like this when the raindrops are quite small and of consistent size. They are predominantly colored purple, pink, and green.

ABOVE Rainbow with supernumeraries, spotted over Valley of the Gods, Utah, US, by Paul Martini (Member 27,060).

IN THIS NIGHTLIFE PANORAMA, viewed from the window of the International Space Station, the glare of the East Coast cities that never sleep is accompanied by a faint red line high above. This is the airglow of the upper atmosphere, a subtle canopy of light up at altitudes of 150–300 kilometers (90–190 miles). Though it has more than one source, a main cause of the glow is energy released by oxygen atoms still excited from the sunlight of the day before.

ABOVE Airglow in the upper atmosphere, spotted by astronaut Mark Vande Hei aboard the International Space Station.

ABOVE Someone, bottom right, places a call to the local authorities to report a pothole in the sky. Spotted in Stratocumulus clouds over Newton, Massachusetts, US, by Fiona Graeme-Cook (Member 44,036).

OPPOSITE Detail from frontispiece of *An invective against Cathedral Churches, Church-Steeples, Bells, etc.* (1656), by Samuel Chidley.

ON OCTOBER 21, 1638, a thunderstorm raged over the village of Widecombe-in-the-Moor in Devon, England. A service was being held in the local church at the time, which was interrupted by what the reverend later described as a "great fiery ball" entering through a window. The strange apparition caused chaos inside, killing four members of the congregation and injuring 62 others. It was one of the earliest recorded incidences of ball lightning, a phenomenon associated with thunderstorms that has been reported many times since but is still little understood. Unpredictable and transient, it has been practically impossible to study, and so there are several competing theories for its cause. To this day, the "great fiery ball" that caused so much havoc in a Devon church in the 17th century remains one of the unsolved mysteries of nature.

THIS COFFIN SHAPE beneath a layer of marine Stratocumulus clouds is in fact an iceberg spotted from the International Space Station. The iceberg is a large fragment of a parent berg known as B-15 that broke off the Ross Ice Shelf of Antarctica in March 2000. For over 18 years, the iceberg has floated around the Southern Ocean as collisions have cleaved off parts like this one. The coffin-shaped iceberg has now floated up into the South Atlantic, between the island of South Georgia and the South Sandwich Islands. These Atlantic waters are warmer than those of the Southern Ocean and so B-15T, as the fragment is known, has finally made its way to the place where icebergs go to die.

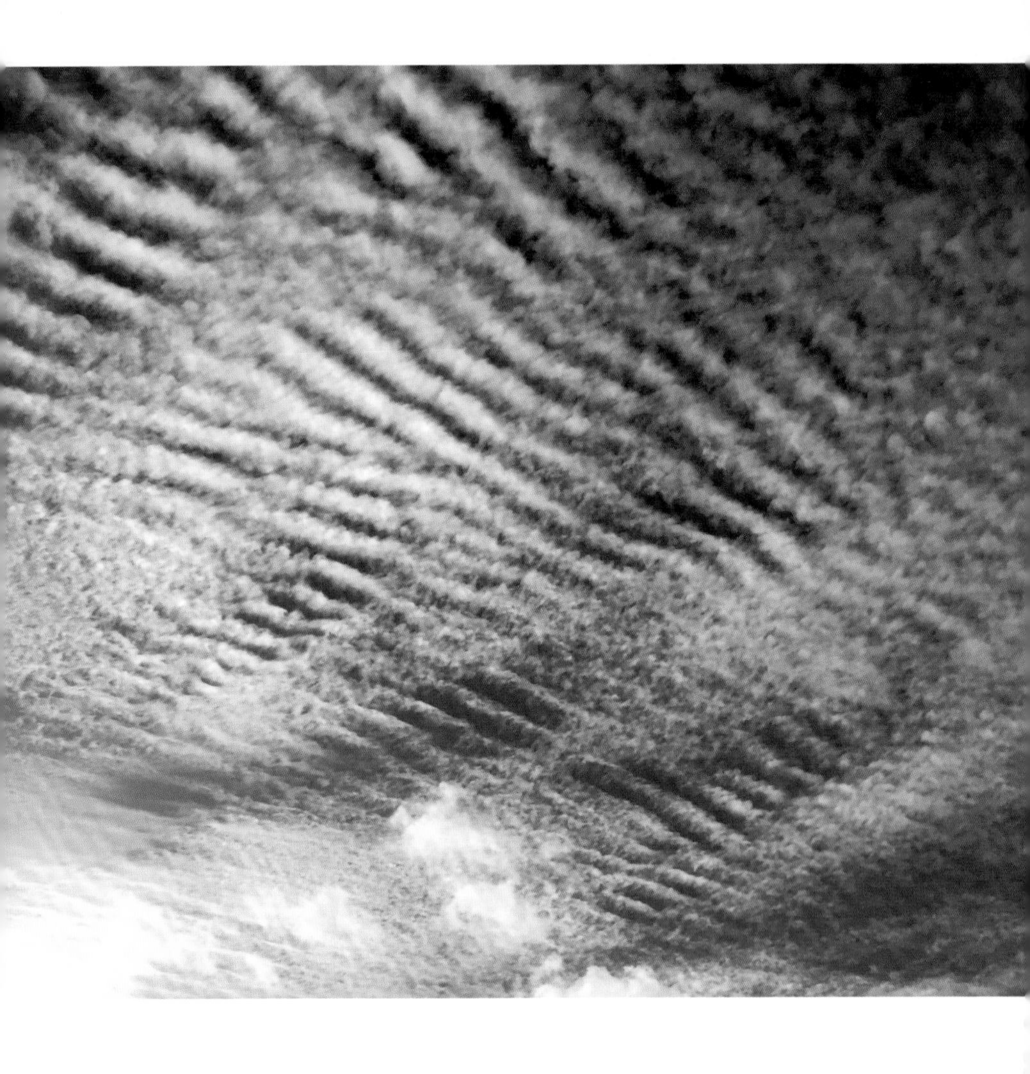

OPPOSITE The B-15T iceberg, laid to rest beneath a quilt of Stratocumulus over the South Atlantic, spotted by a member of the Expedition 56 crew aboard the International Space Station.

ABOVE Altocumulus undulatus clouds spotted over Knox, Victoria, Australia, by Nicole Bates (Member 38,201). Like the ridges of sand beneath your feet as you walk on the beach at the water's edge, except that they are made of cloud, not sand. And they are high above, not down below. And they form in a completely different way. Otherwise, very similar.

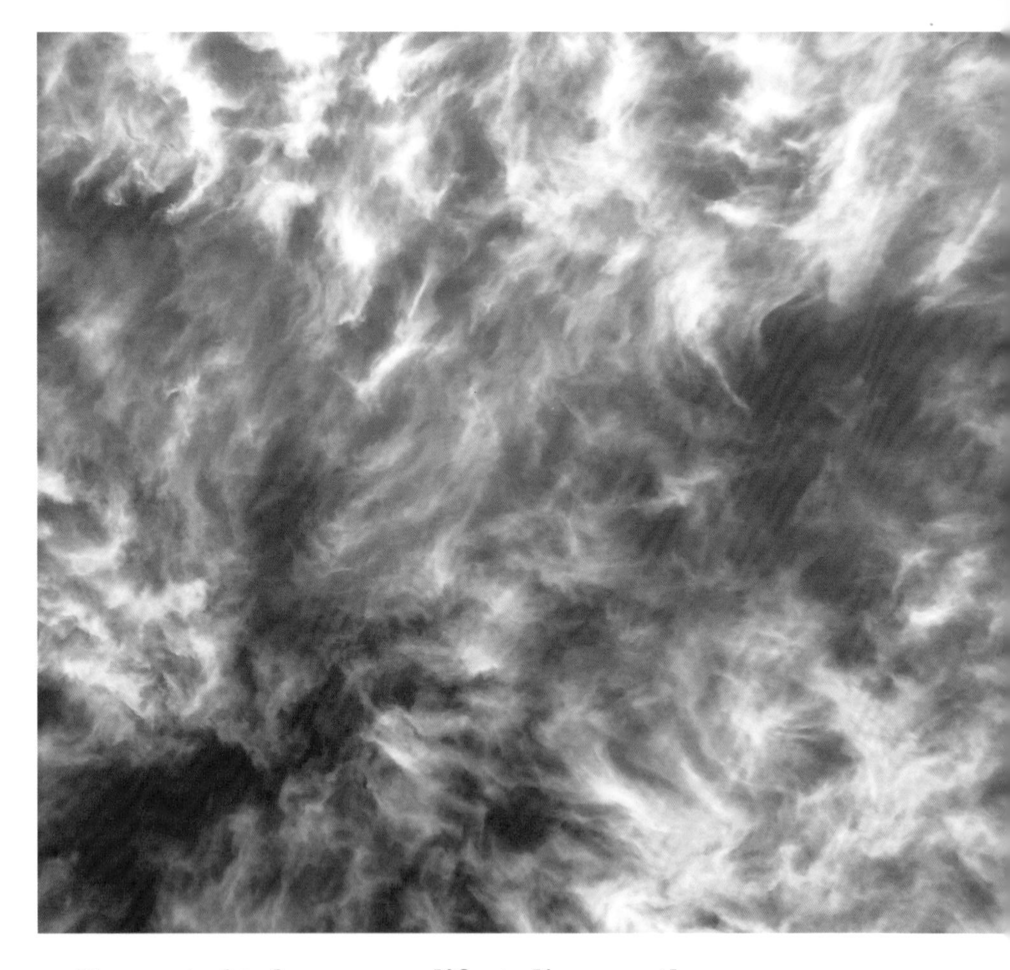

He wanted to have some life; to lie upon the earth, and smell it: to look up into the sky like anthropos, and lose himself in clouds. He knew suddenly that nobody, living upon the remotest, most barren crag in the ocean, could complain of a dull landscape so long as he would lift his eyes.

From *The Book of Merlyn* (1977), by T. H. White.

ABOVE Cirrus intortus, spotted over Napa, California, US, by Patricia "Keelin" (Member 41,642).

OPPOSITE Cumulus fractus cataractagenitus, spotted over a waterfall near Ásólfsskáli, Iceland, by Enrique Roldán (Member 12,510).

THE LATIN NAMING SYSTEM for clouds refers mostly to their appearance and altitudes. But there are some classification terms that relate to how the cloud came into being. These end with -genitus, which means "made." One example is the snappy term of cataractagenitus, used for clouds that are made by a waterfall. Tall waterfalls, like this one near Ásólfsskáli in Iceland, can cause the air to become saturated with moisture. The relentless cascade of water also drags air downwards, causing neighboring regions of air to lift up and replace it. Whenever saturated air lifts, cooling a little as it does so, there is the chance for cloud to form. In this image, the waterfall-generated cloud is the pale, frayed-looking Cumulus just above the mouth of the waterfall, not the bright white Cumulus higher in the sky.

THE NEW ASPERITAS CLOUD FEATURE, which was first identified by members of the Cloud Appreciation Society, can appear in Stratocumulus and Altocumulus layers. The turbulent and chaotic waves of the rare formation appear mostly in the vicinity of storms. The classification became official in 2017, thanks to photographs taken not by official weather observers but everyday people who happened to notice the sky when out and about—in the parking lot, for instance. Enabled by mobile technology, this new distributed perspective on the sky might eventually turn out to be as revolutionary for our view on the atmosphere as that provided by satellites.

ABOVE Asperitas in Stratocumulus clouds, spotted by Daisy Dawson over Southwick, East Sussex, England.

THIS DOUBLE RAINBOW looks as if it is pierced by rays of light. The effect is known as a rainbow wheel because the rays can occasionally appear right around the whole arc of the bow. The rays converge due to the effect of perspective as they recede off towards the horizon. They are a form of anti-crepuscular rays, shadows cast from tall Cumulus clouds in front of a Sun that is behind the viewer. Perspective means that they fan out from the anti-solar point, which is in the opposite direction from that of the Sun. Since a rainbow is centered on the anti-solar point too, the rays appear within it like spokes of a bicycle wheel. A very hippie-looking bicycle wheel.

ABOVE Rainbow wheel, spotted over Merewether Beach, New South Wales, Australia, by Elizabeth Freihaut (Member 40,658).

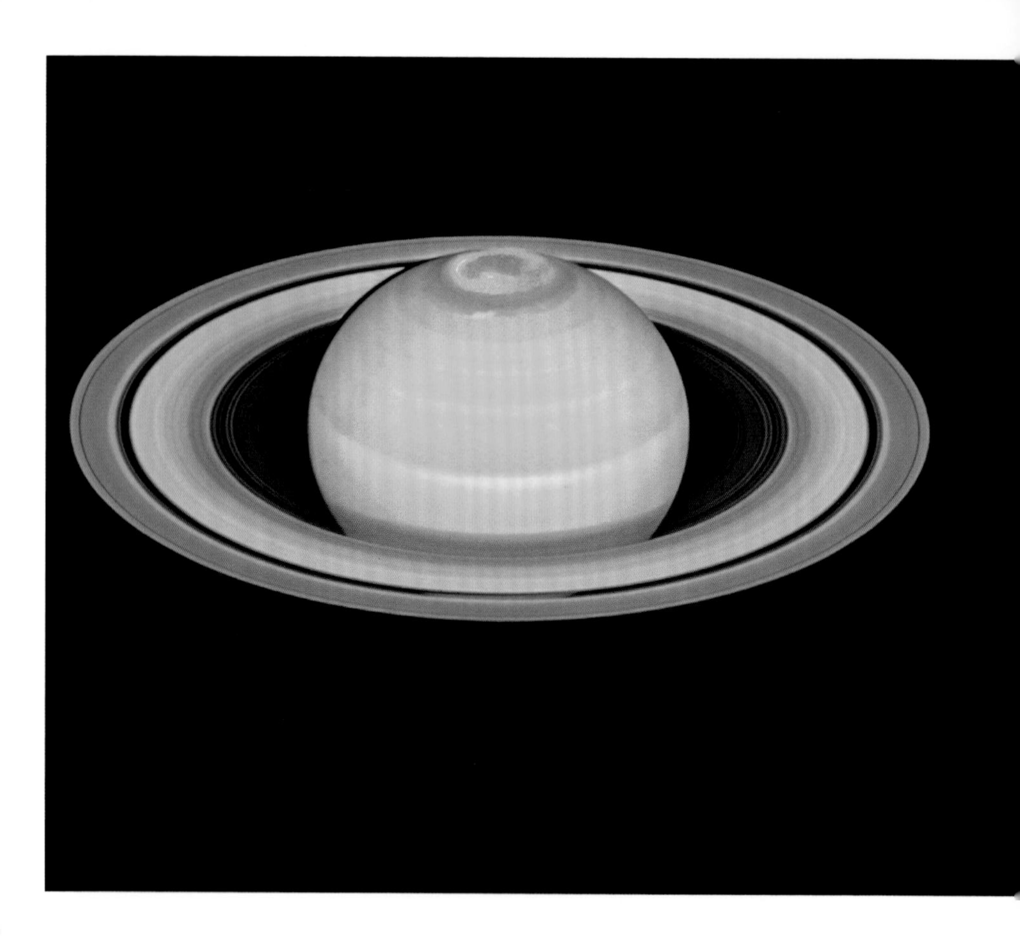

EARTH IS NOT THE ONLY PLANET in our solar system to produce auroras. They also appear on the four gas giants, including Saturn. Auroras are created when charged particles emitted from the Sun as solar wind become caught in a planet's magnetosphere before being drawn towards its magnetic poles. These particles excite gas atoms and molecules in the upper atmosphere and cause them to emit light. Saturn's atmosphere has lots of hydrogen, which means that its auroras are largely in the ultraviolet spectrum, rather than the visible. Saturn's auroras are hidden to telescopes on Earth because our atmosphere filters out much of the ultraviolet wavelengths of light. Fortunately, floating high in the clarity of Space, the Hubble Space Telescope can capture the ultraviolet aurora, to reveal the lights that dance around Saturn's distant poles.

ABOVE Hubble Space Telescope image of Saturn's ultraviolet aurora superimposed onto a visible-spectrum image of the planet.

WITH AN AREA OF 10,582 square kilometers (4,086 square miles), Bolivia's Salar de Uyuni is the largest region of salt flats in the world. After the wet season, parts can become covered in a layer of rainwater just a few centimeters deep. No waves can develop in such shallow water, making the Salar de Uyuni an enormous mirror on the sky. This is Cloud Cuckoo Land.

ABOVE Cumulus and lenticularis reflections on the Salar de Uyuni, Bolivia.

There was a fecklessness, a lack of symmetry and order in the clouds, as they thinned and thickened. Was it their own law, or no law, they obeyed? Some were wisps of white hair merely. One, high up, very distant, had hardened to golden alabaster; was made of immortal marble. Beyond that, was blue, pure blue, black blue; blue that had never filtered down; that had escaped registration.

From *Between the Acts* (1941), by Virginia Woolf.

ABOVE Stratocumulus, Cumulonimbus, and velum, spotted by Henrik Välimäki over Lapua, Finland.

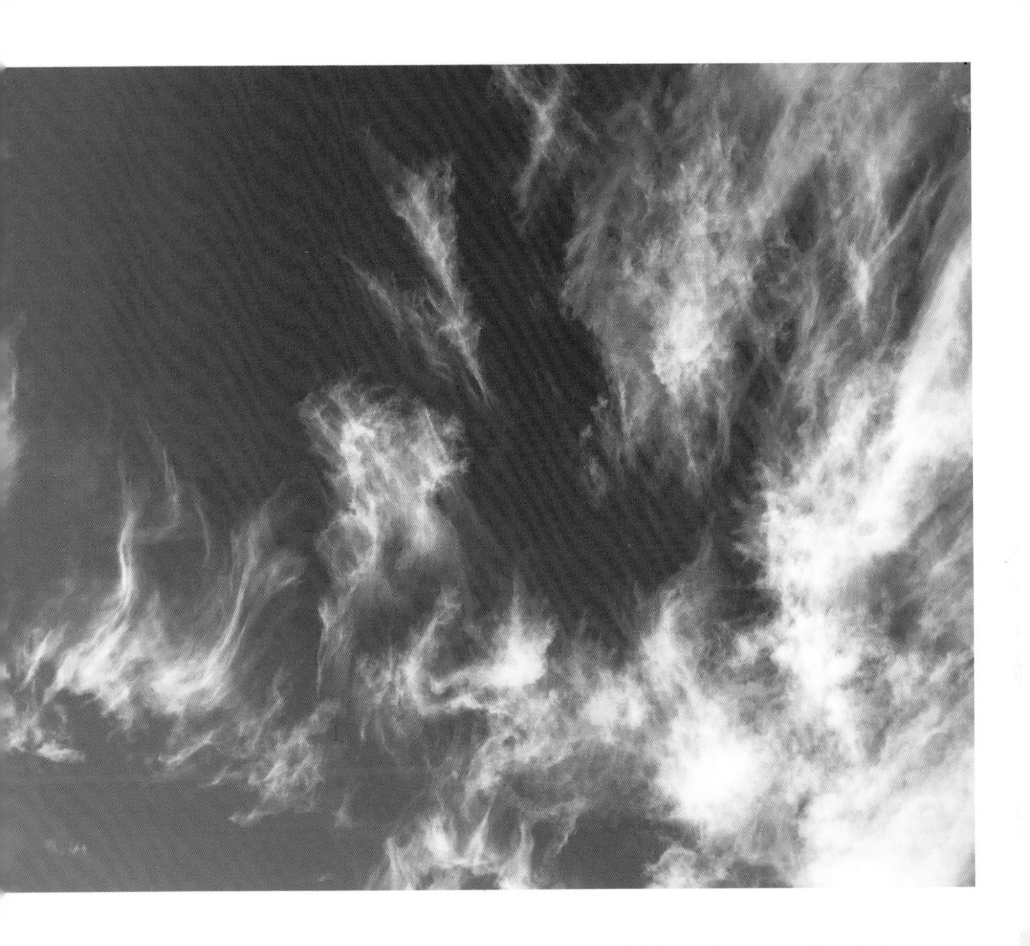

CIRRUS IS ONE OF THE ten main cloud types, which are known collectively as the cloud genera. As with other genera, particular examples of Cirrus can often be described more specifically by adding one or more terms, known as cloud species and varieties. The ones used for Cirrus clouds tend to describe patterns in their high, ice-crystal streaks. Intortus, the variety shown here, is named from the Latin for "twisted" or "tangled." Unlike the more orderly patterns like fibratus (streaks aligned in long parallel filaments), uncinus (extending from hooked ends), and floccus (arrayed in fluffy tufts) the intortus variety is when the Cirrus wisps are randomly oriented, turning this way and that—twisted into confusion by the capricious currents of the upper troposphere.

ABOVE Cirrus intortus, spotted by Peter van de Bult over Oranjewoud, Heerenveen, Friesland, Netherlands.

ABOVE An angel with a camcorder, spotted over the Kennet and Avon canal, Wiltshire, England, by Anne Hatton (Member 14,125).

OPPOSITE Sea smoke, also known as steam fog, at sunrise over South Portland, Maine, US, spotted by Margaret D. Webster (Member 40,825).

SUNRISE OVER THE SNOWY COAST of Maine, US, reveals the Atlantic Ocean topped with a layer of sea smoke. This variety of fog is caused by very cold air flowing over a body of comparatively warm water. Air just above the surface is heated by its contact with the surface and picks up moisture evaporating off the water. As this air mixes with the gentle flow of cold air above, it cools rapidly and some of the moisture it picked up condenses into visible droplets. The effect is rather like the steam rising from a mug of tea in a cold room. Usually, sea smoke is thin, rising from the surface in delicate columns that disperse quickly in the breeze. Occasionally, like on this particular morning, it can be much denser. Sea smoke's rising columns of condensation have even been known to reach as high as 20 meters (65 feet).

THE WHITENESS OR GREYNESS of a cloud depends on a number of factors. One is whether the side you are looking at is lit by the Sun or in shadow. Here, the Sun is ahead and to the right. The large Cumulonimbus cloud in the distance looks dark because the light is shining onto it from behind, so we are seeing its shaded side. Another factor is whether a cloud is made of droplets or ice crystals. A low Cumulus cloud consists of a great number of tiny droplets of water. This makes it look more solid than a high Cirrus, which consists of fewer and larger ice crystals. The low cloud's droplets scatter the light more than the less plentiful ice crystals of the high cloud, and so Cumulus look bright white as they reflect the sunlight, while Cirrus look paler as they let more light through. While the Cumulus in the foreground here are lit from behind, they are thin because they're breaking up and so more of the light shines through. One more factor has to do with our own perception. We judge light and shade in a relative way. Against a dark background like this, we see these little Cumulus as bright, while against the brilliance of a clear sky they'd appear darker.

OPPOSITE Cumulus in the foreground with Cumulonimbus in the background, spotted by Sofie Bonte over Holsbeek, Belgium.

UNLIKE THE EPHEMERAL, ever-shifting atmospheric clouds, the cloud-like form of the Milky Way is unchanging and eternal as it stretches out across our night skies. This is not, of course, a cloud of droplets scattering light from our Sun, but instead the collective light from all the other suns arrayed across our galaxy. These likely number around a couple of hundred billion. The droplets within a typical large Cumulus cloud might amount to the same number.

ABOVE The Milky Way, spotted from Bluff, Utah, US, by Paul Martini (Member 27,060) as lightning strikes on the horizon.

OPPOSITE A 22-degree lunar halo, spotted by Doug Short (Member 21,013) over Anchorage, Alaska, US.

THE SUN CAN PRODUCE a whole range of optical effects as its light interacts with water in the atmosphere. Less common, and even more magical, are the optical phenomena caused by the Moon. Its light can produce all the same effects as the Sun's, but lunar bows and halo phenomena are observed far less frequently because they are usually too faint for our eyes to pick out. Except, of course, when the Moon is full. At its brightest, when the sky is wreathed in an ice-crystal layer of Cirrostratus, the Moon can trace a broad ring of light known as a 22-degree lunar halo, as its silvery glow refracts through the cloud's tumbling prisms of ice.

CIRROSTRATUS IS THE MOST UNDERSTATED of the ten main cloud types. It is a high layer of ice crystals that often covers the whole sky; just a faint, featureless, milky whitening of the blue. In fact, when a layer of Cirrostratus is thin, many would consider there to be no cloud present at all, remarking that the sky is merely a little paler than normal. Only when it thickens does Cirrostratus start to become more noticeable, like the example spotted by Danish artist Christen Købke in this 1830s painting of the view from the roof of Frederiksborg Castle in Hillerød, Denmark. Such a retiring cloud prefers not to remain in the limelight for long. No sooner has it grown dense enough to be remarked upon than the Cirrostratus's base usually will have descended sufficiently for us to reclassify it as the mid-level layer cloud, Altostratus.

EMILY WATSON SPOTTED this fogbow on vacation in Tenerife. With a low Sun shining from behind her and a delicate shroud of mist up ahead on the slopes of Mount Teide, the island's extinct volcano, she knew the moment was right for a fogbow. Emily headed to a vantage point from where she could look down the line of the sunlight as it shone onto the cloud below. Fogbows, or cloudbows, are like rainbows produced by droplets far smaller than raindrops. Typically, they are observed looking down onto fog from a mountainside or onto higher cloud from an aircraft. Also just visible here is a faint ring of colors around Emily's shadow in the center of the bow. This is the related optical effect known as a glory. The fogbow effect was a special sight on what turned out to be a special vacation, on which Emily became engaged to marry.

ABOVE Fogbow, spotted by Emily Watson (Member 44,218) in Tenerife, Canary Islands.

OPPOSITE *Roof Ridge of Frederiksborg Castle with View of Lake, Town and Forest* (1833–1834), by Christen Købke.

THE POCKETS OF RISING AIR known as convection cells that form Altocumulus and Cirrocumulus clouds into regular patterns of cloudlets with gaps between can lead more rarely to the exact opposite pattern: a regular array of holes separated by fringes of cloud. This is the variety known as lacunosus, when the convection is described as having open cells, in contrast to the closed cells that lead to cloudlets. The holes appear where denser, cooler air is sinking from above, while the fringes of cloud are where less dense, warmer air is floating up from below to replace it. Lacunosus is often described as resembling a honeycomb. This one looks more like the ragged sail of a galleon shot through with cannon fire. What's the Latin for that?

All other creatures look down toward the earth, but man was given a face so that he might turn his eyes toward the stars and his gaze upon the sky.

From *Metamorphoses* (8 AD), by the Roman poet Ovid.

OPPOSITE Altocumulus lacunosus, spotted over Owens Valley, California, US, by Stephen Ingram (Member 7,328).

ABOVE Cumulus cloud caught by Maria Lyle over Sarasota, Florida, US.

MOST TREES OBTAIN the water they need through their roots, either from rainwater or groundwater. But studies in the coastal redwood forests of California, US, and the tropical forests of Monteverde, Costa Rica, have shown that some trees absorb water more directly as their leaves become drenched with droplets of fog. It is a process called foliar uptake and it comes in handy for trees in forests that experience a dry season with minimal rain but abundant fog. Whether this fog drifts into coastal forests from the open ocean or it rises up the wooded slopes of mountains inland, the fine hairs on the leaves of these trees allow them to snag the tiny fog droplets to cut out the middle man and extract water directly from the clouds.

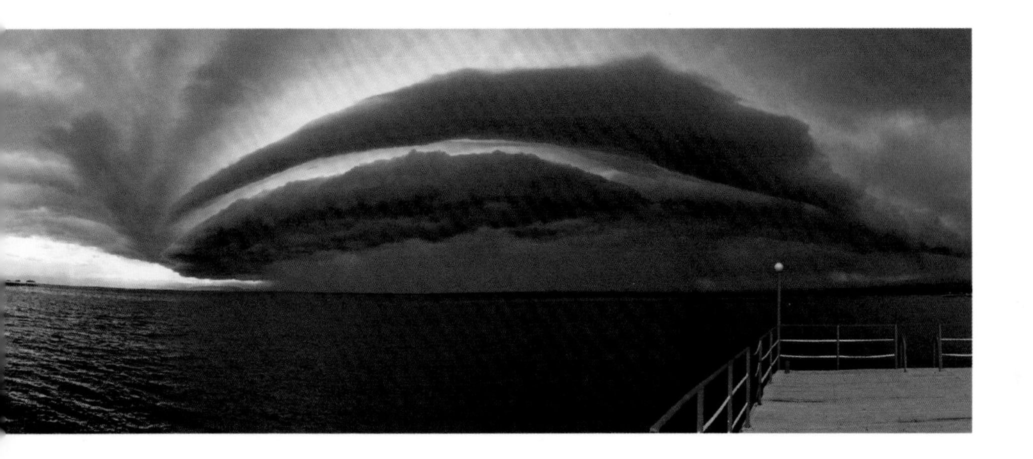

YOU'LL KNOW A SERIOUS STORM is on its way when you see along its front base the dramatic cloud feature known as an arcus. It is also called a shelf cloud, and it appears at the front of an advancing supercell storm. The arcus acts as a last-minute warning that wild weather is imminent. It is not always obvious, however, why the feature should be named with the Latin word for "arch" or "vault." Not unless you see a particularly large example. Then, the effect of perspective will make the low ridge of cloud appear to bulge upwards where its middle is nearest and curve downwards where its edges recede. This gives it the appearance of a dark, ominous arch.

OPPOSITE Stratus-filled peaks over Urubici, Brazil, spotted by Roberval Santos (Member 24,490).

ABOVE Arcus, or shelf cloud, spotted over Issyk-Kol, eastern Kyrgyzstan, by Busra Karademir (Member 45,062).

THIS JANUARY FOG is following the river along the foot of Mount Eglinton, New Zealand. As air rises off the mountain's Sun-warmed slopes, cooler air is drawn across the valley floor from the shaded side to replace it. Passing over Eglinton River, this cool air mixes with the evaporation lifting off the warm summer waters to form a low line of fog known as river smoke.

ABOVE River smoke fog, spotted from Eglinton Flats, South Island, New Zealand, by Jean Gray.

AN UPSIDE-DOWN HORSESHOE is considered unlucky by some, but not cloudspotters. The horseshoe vortex cloud is a rare prize to add to your collection of cloud formations. The delicate curve of the horseshoe shape twists in the wind as its center lifts upwards through the sky. Not only is the cloud rare, it is also fleeting. After just a minute or two it will have evaporated away, soon nothing more than a memory for whoever was lucky enough to have noticed it.

ABOVE Horseshoe vortex cloud, spotted by Stephanie Arena over Cazenovia, Wisconsin, US.

TOP The Cat's Paw Nebula, spotted by NASA's Spitzer Space Telescope. This gargantuan interstellar cloud is blowing bubbles as the heat from newborn stars causes the gas around them to expand. The bubbles will eventually burst to form a honeycomb-like structure.

ABOVE Like those tiny fish that swim beside sharks, velum clouds are easily missed as they form at the flanks of show-stealing Cumulonimbus. The subtle dark streaks were spotted by Fiona Semmens over Huskisson, Australia.

OPPOSITE Altocumulus beneath patches of Cirrostratus, spotted at sunset by Marty Bell (Member 46,529) over Manasota Key, Florida, US.

Gaze into the fire, into the clouds, and as soon as the inner voices begin to speak . . . surrender to them. Don't ask first whether it's permitted, or would please your teachers or father or some god. You will ruin yourself if you do that.

From *Demian: Die Geschichte von Emil Sinclairs Jugend* (1919), by Hermann Hesse.

ABOVE Pi in the sky, spotted by Patty Kjobmand Cashman over Mount Shasta, Northern California, US.

OPPOSITE *The Translation of the Holy House of Loreto* (mid-1490s), attributed to Saturnino Gatti.

Whereas the medieval never painted a cloud but with the purpose of placing an angel upon it . . . we have no belief that the clouds contain more than so many inches of rain or hail.

From *Modern Painters* (1856), Volume III, Chapter 16, by John Ruskin.

CUMULUS CLOUDS FORM ON COLUMNS of air rising off the Sun-warmed ground. The upward currents of convection, known as thermals, expand as they rise, which makes the air cool. Sometimes it cools enough for the moisture it contains to condense into droplets we see as cloud. For this reason, Cumulus are like beacons of the sky, revealing the shifting locations of these invisible elevators of air. Glider pilots ride thermals to gain lift. It is a trick we have learned from birds like these gulls. They use the thermals to help them migrate over vast distances while expending minimal energy, gliding from one thermal to the next with barely a flap of their wings. Do they feel the proximity of the thermal updrafts with the finest feathers at their wingtips? Or are they cloudspotters, like their pilot companions, who find them by reading the Cumulus?

I have seen the softness and beauty of the summer clouds floating feathery overhead, enjoying, as it seemed, their height and privilege of motion.

From "Essay VI: Nature," *Essays* (1906), by Ralph Waldo Emerson.

OPPOSITE Gulls riding the thermals of Cumulus clouds, spotted by Peter Dayson (Member 27,411) over Burry Port, Wales.

ABOVE Cirrus, spotted by Marc van Workum over Stocken-Höfen, Switzerland.

CLOUDS CAN TRICK THE EYE with their play of light and shade. Here, the outline of a tall Cumulus appears to have been cut from a thin layer of Altocumulus above. In fact, the Cumulus tower extends up through the Altocumulus layer. The "cutout" is its shadow cast down onto the layer below. We see cloud droplets because they scatter sunlight in the sky. When they are in a thin layer like this, they appear as a bright blanket in the direct sunlight and almost transparent in the shade. This, combined with the counterintuitive effects of shadows in perspective, reminds us that the architecture of clouds is never quite as it seems.

FOR CITY DWELLERS, the sky is the last true wilderness.

OPPOSITE Cloud shadow cast by Cumulus onto Altocumulus, spotted by David Watson (Member 44,951) over Burgess Hill, West Sussex, England.

ABOVE Altocumulus, spotted over Rose Hill, Manhattan, New York, US, by Tony Hoffman (Member 34,316).

We are the first generation to see the clouds from both sides. What a privilege! First people dreamed upward. Now they dream both upward and downward. This is bound to change something.

From *Henderson the Rain King* (1959), by Saul Bellow, and the passage that singer-songwriter Joni Mitchell referenced as inspiration for her 1967 hit "Both Sides, Now."

ABOVE Altocumulus undulatus from the other side, spotted over the American prairies by Shotsy Faust (Member 45,472).

OPPOSITE Circumzenithal arc and section of a supralateral arc, spotted by Anthony Skellern (Member 19,011) over Bollington, Cheshire, England.

THE UPTURNED BOW of the ice-crystal optical effect called a circumzenithal arc is infrequent but not all that rare. According to Claudia and Wolfgang Hinz of the German halo observation network Arbeitskreis Meteore, circumzenithal arcs form over mainland Europe six times a year on average. What makes this one special is that it is joined tangentially by a fainter and much rarer light effect. Called a supralateral arc, this is the far broader bow that is curving downwards. This one only tends to form about once a year. Both effects are found high in the sky when the Sun is down towards the horizon. You'd only notice them if you chose to look directly above your head, so they are treasures hidden in plain sight from all but those with their heads in the clouds.

WHILE MOST LANDSCAPES painted around the start of the 20th century were depicted from the usual ground-level perspective, this 1906 scene of Cumulus clouds by Russian-born artist Boris Anisfeld was depicted from on high. To achieve this view, he climbed to the top of Aiou-Dagh Mountain on the Crimean Peninsula.

ABOVE *Clouds over the Black Sea—Crimea* (1906), by Boris Anisfeld.

OPPOSITE Hanggliders, turn left ahead. Spotted by Nienke Lantman (Member 24,009) over Erm, Netherlands.

JOHAN CHRISTIAN DAHL was a 19th-century Norwegian painter considered a leading light of the Romantic period. He decided early in his career to study "nature above all," and so his subjects were invariably land- and seascapes. His aim was to create art "representing nature in all its freedom and wildness." No wonder he chose to let the majority of his canvases be dominated by the sky.

ABOVE Detail from *Clouds and Sunbeams over the Windberg near Dresden* (1857), by Johan Christian Dahl.

CREPUSCULAR RAYS caused by the shadows cast from clouds before the setting Sun, shown on the left, can sometimes stretch right overhead and all the way to the opposite horizon, shown on the right. The rays visible when facing away from the Sun like this are known as anti-crepuscular rays. Stretching right across the sky, the straight rays appear to bulge out directly above and converge near the horizons due to perspective. It's hard to show in a photograph. You kind of have to be there—as Mary Stivison was when she spotted these at sunset.

ABOVE Crepuscular rays, above left, and anti-crepuscular rays, above right, spotted by Mary Stivison looking towards and away from the sunset over Clarksville, Johnson County, Arkansas, US.

It is a very beautiful day. The woman looks around and thinks: "there cannot ever have been a spring more beautiful than this. I did not know until now that clouds could be like this. I did not know that the sky is the sea and that clouds are the souls of happy ships, sunk long ago. I did not know that the wind could be tender, like hands as they caress—what did I know—until now?"

From Dark Spring (1954), by Unica Zürn. English translation from the German by Caroline Rupprecht (2000).

ABOVE Cumulus, spotted over Tempe, Arizona, US, by Laura Simms (Member 32,141).

OPPOSITE Detail from Piero della Francesca's *Burial of the Sacred Wood* fresco (c. 1452–66) in the Basilica of San Francesco, Arezzo, Italy.

THE EARLY RENAISSANCE PAINTER Piero della Francesca was clearly a cloudspotter. He filled his skies not with the typical cartoon-like Cumulus that everyone else painted around this time, but with distinctive lenticularis clouds. These are the ones that look rather like flying saucers. They appear in his famous frescoes in Arezzo as well as in several other paintings. The fact that they form downwind of hills and mountains might explain why Della Francesca was so fond of them. He was born and raised in the Apennine Mountains of Tuscany. Perhaps he gazed up at them as a youngster. Perhaps these distinctive clouds left an impression on him that would influence the skies he would paint later in life.

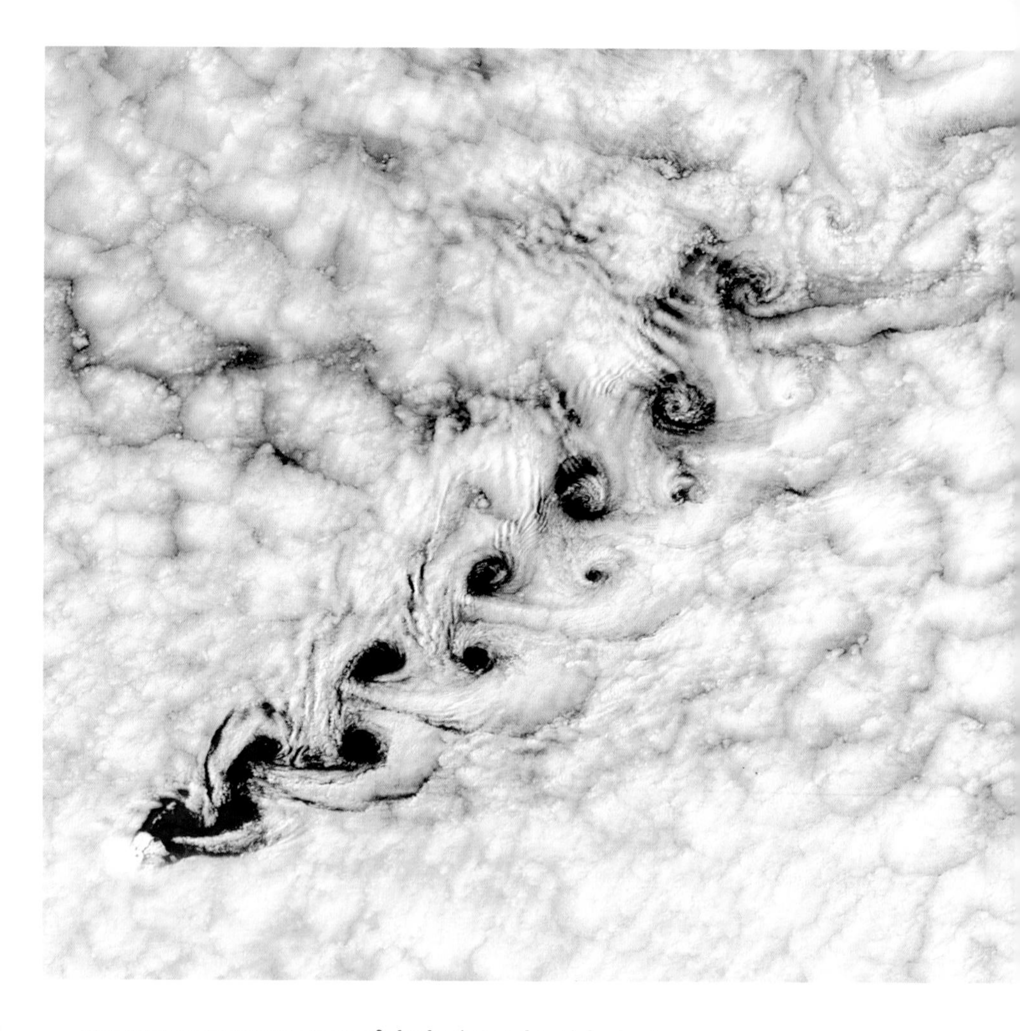

THE SWIRLING SUCCESSION of clockwise and anticlockwise eddies, known as Von Kármán vortices, form in the maritime cloud cover downwind of Heard Island in the south Indian Ocean. It's nature's lesson in braiding hair.

ABOVE Von Kármán vortices, spotted over the south Indian Ocean by NASA's Terra satellite.

CLOSE YOUR EYES, think of a cloud, and it is probably a Cumulus that comes to mind. These are the generic ones, the iconic formations, and they form on top of thermals, columns of air rising invisibly off the sun-warmed ground. When they are small, fair-weather examples like these over the Simpson Desert in Australia, they are known as Cumulus humilis. Aah.

ABOVE Cumulus humilis, spotted by Sinead Hurley over the Simpson Desert, south-east of Alice Springs, Australia.

THE UDDER-LIKE FEATURES of mamma clouds exist to remind us that our lives depend upon the milk of clouds. They are the planet's great recyclers, expressing the endless flow from evaporation to condensation to precipitation that turns water from saline to fresh.

ABOVE Mamma formed beneath the canopy, or incus, of a Cumulonimbus storm cloud, spotted over Drayton, Oxfordshire, England, by Lucy Cuckney (Member 14,259).

OPPOSITE Nimbostratus, spotted over Holme, Aarhus Kommune, Jutland, Denmark, by Søren Hauge (Member 33,981).

Mountains and oceans have whole worlds of innumerable wondrous features. We should understand that it is not only our distant surroundings that are like this, but even what is right here, even a single drop of water.

From *Genjōkōan* (1233), by Dogen Zenji, a Japanese Buddhist priest.

ABOVE A UFO in stealth mode behind Mont Blanc, France. Spotted from Les Chosalets, Chamonix, by Eystein Mack Alnæs. Also known as Altocumulus lenticularis.

OPPOSITE, LEFT TO RIGHT Spotted over Queens, New York, US, by George Preoteasa (Member 41,445); spotted over Nantwich, England, by Jan McIntyre (Member 34,229); spotted over Hérouville-Saint-Clair, France, by Thibaut de Jaegher.

THREE CLOUD VARIETIES describe how transparent a cloud is. A layer cloud like the Altostratus on the left is called “translucidus” when it is thin enough for you to be able to see the position of the Sun (hmm). One like the Stratus layer in the middle is called “opacus” when it is thick enough to completely obscure the position of the Sun (boo). And a cloud like the Altocumulus on the right is described as “perlucidus” when it has thicker clumps with gaps between which the upper sky shows through (yay!).

ABOVE The fiery light from a low Sun illuminates curtains of evaporating precipitation, known as virga, falling from Altocumulus clouds through the skies of Florianopolis, Brazil. Spotted by Roberval Santos (Member 24,490).

OPPOSITE A shock-wave cloud engulfs a Typhoon jet, spotted by Ross McLaughlin over Portrush, Northern Ireland.

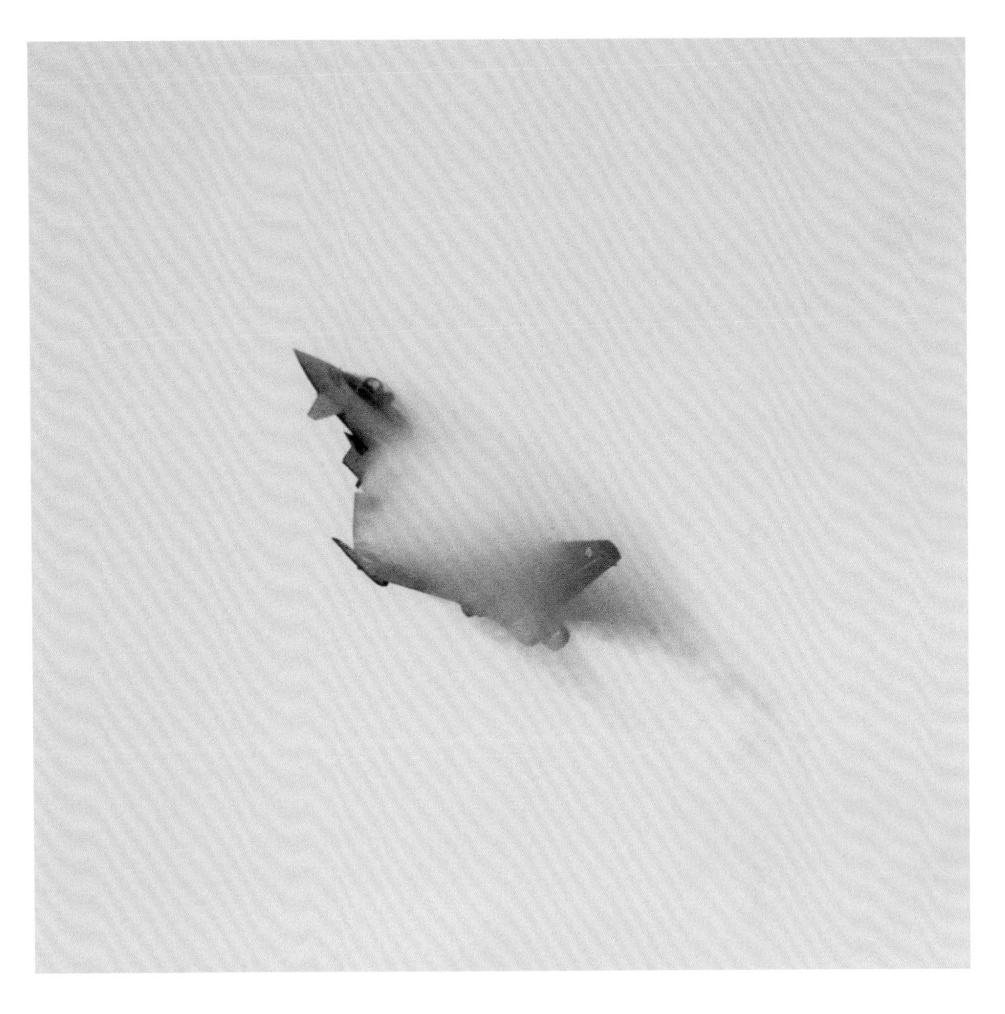

A JET APPROACHING SUPERSONIC SPEEDS can create a shock wave in the air it passes through. You can think of the highly compressed air at the plane's nose like a crest, or peak, in this air-pressure wave. The crest of any wave is followed by a trough, and so a corresponding region of extremely low air pressure forms a short distance behind it. This is where a cloud can form and envelop the jet. Depending on its shape, the shock-wave cloud is known sometimes as a vapor cone or a shock egg. The droplets form because a rapid drop in air pressure means an equally rapid drop in temperature, which can cause the air's moisture to condense into liquid. Each droplet appears for just a fraction of a second before warming and evaporating again in the aircraft's wake. The pressure wave stays fixed in position relative to the jet, so the cloud clings to its speeding fuselage.

THIS MURAL IS IN THE LOBBY of the Mesa Laboratory in the National Center for Atmospheric Research (NCAR) in Colorado. It was conceived in 1974 by NCAR atmospheric scientist Melvyn Shapiro and executed with artist Howard Crosslen. It incorporates a storm cloud spawning a tornado, fluctus cloud formations, mountain airflows, and even the prow of a Viking longship—to honor the Scandinavian scientists who contributed so much to 20th-century meteorology. The mural symbolizes our dynamic atmosphere and the forces that drive it.

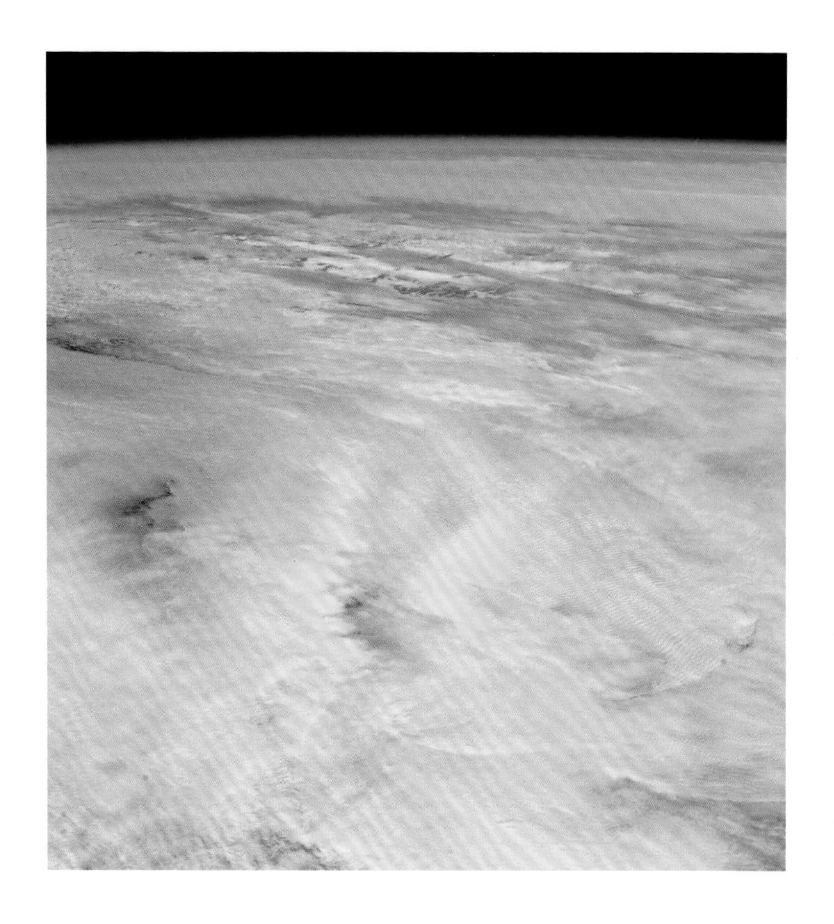

FLYING CLOUDSPOTTERS will be familiar with the glory optical effect that can sometimes be spotted from the window seat. It appears as a ring of rainbow colors around the aircraft's shadow as it is cast down onto a layer of cloud below. The effect is formed by sunlight diffracting around the tiny droplets as it reflects off the cloud. And while a glory is noticed most often when the aircraft is close to the surface of a cloud, it can appear when it is flying much higher in the sky. In fact, as this view from the window of the International Space Station shows, a glory can even be seen when flying so high as to be outside of the sky.

ABOVE Glory produced by Altostratus and Altocumulus, spotted by astronaut Alexander Gerst from aboard the International Space Station.

OPPOSITE Mural by Howard Crosslen, spotted at the National Center for Atmospheric Research, Boulder, Colorado, US.

THE GLITTERING DIAMOND DUST ice fog that formed this sun dog optical effect develops naturally in fairly dry and very cold conditions, when the scant water vapor in the air freezes directly into ice crystals rather than condensing first as water. But it can also form as a byproduct of the snow machines used on ski slopes. The minuscule, hexagonal ice crystals, formed by the machines overnight, drift around skiers as they take to the slopes in the chilled air of morning. The crystals produced in this way tend to be very regularly shaped and optically pure. This makes them ideal as the tiny floating prisms of ice that reflect and bend the sunlight to produce halo phenomena like this sun dog.

Perhaps I am just a hopeless rationalist, but isn't fascination as comforting as solace? Isn't nature immeasurably more interesting for its complexities and its lack of conformity to our hopes? Isn't curiosity as wondrously and fundamentally human as compassion?

From "Tires to sandals," *Eight Little Piggies: Reflections on Natural History* (1993), by Stephen Jay Gould.

ABOVE Altocumulus at sunset, spotted by Renee Gerber over Yellowknife, Northwest Territories, Canada.

OPPOSITE Sun dog, spotted over Arc 1600, Les Arcs, France, by Jammin Palmer (Member 46,679).

WHEN CLOUDS FORM during the day over land in preference to over the water, it is generally because the surface of the land heats up more rapidly in the sunshine than the water does. But in the case of an enormous tropical rainforest like the Amazon there is another reason: the trees. During the dry season in particular, the trees release moisture into the air through their leaves as they are heated by the Sun. This is like a plant version of sweating known as transpiration, and it can release enough moisture into the air to encourage the formation of clouds like the vast array of fair-weather Cumulus shown here. Studies of satellite data suggest that the trees of the Amazon rainforest are actively involved in shifting the region's weather from dry season to wet. Within the flow of air over the Amazon from the Atlantic to the Andes, more than half of the rain is generated by the forest itself in a repeating cycle of transpiration and precipitation. In fact, the collective effect of all the trees in the Amazon means that the flow of airborne water in the skies above the region is greater throughout the year than that in the enormous river system below.

OPPOSITE Cumulus clouds, spotted over the northwestern Amazon basin in 2009 by NASA's Aqua satellite.

A certain recluse monk once remarked, "I have relinquished all that ties me to the world, but the one thing that still haunts me is the beauty of the sky." I can quite see why he would feel this.

Essays in Idleness (c. 1330), by the medieval Japanese author and Buddhist monk Yoshida Kenkō.

ABOVE Altocumulus, spotted by Deborah Milics over Klemzig, South Australia, Australia.

OPPOSITE A glory appearing in thin Stratus, spotted by Paul Harwood over Mount Jancowski, Coast Mountains, British Columbia, Canada.

OF ALL THE CLOUD OPTICAL EFFECTS, the one known as a glory has to be the most egocentric. Allow us to explain. The effect appears as a ring of rainbow colors when direct sunlight casts a shadow onto cloud, and it is spotted typically when looking onto cloud from an aircraft or a mountain. The egocentric part is that each person in a group observing a glory together sees something different from their neighbor. Each sees the ring of colors centered around their own shadow and none around those of their companions. The image here is a case in point. For the photographer, the delicate rings are centered perfectly on his silhouette, while for the climber ahead it is his own shadow that plays the starring role. Glories are the optical effects for the social media age.

THE TRAVELERS IN HOKUSAI'S PRINT *Mishima Pass in Kai Province* are so excited to have reached an oversized cypress tree that they've completely missed Mount Fuji's glorious banner-cloud display in the distance. Crazy tree-huggers!

ABOVE Detail from *Mishima Pass in Kai Province* (c. 1830–1832), from the series *Thirty-Six Views of Mount Fuji*, by Katsushika Hokusai.

OPPOSITE Asperitas, spotted over Burnie, Tasmania, Australia, by Gary McArthur (Member 5,353).

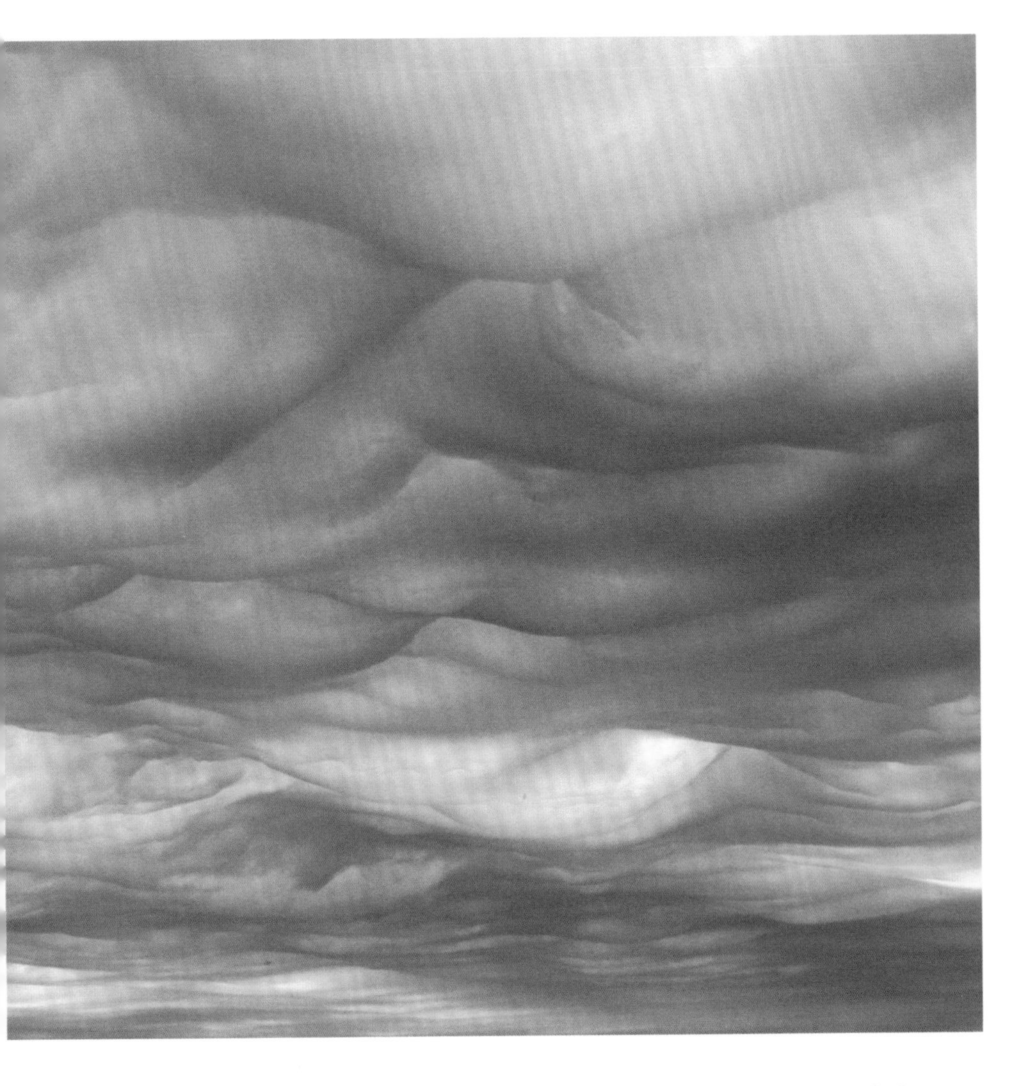

HOW DO YOU DISTINGUISH the wavy form of the recently classified asperitas cloud from the other wave cloud, undulatus? We are glad you asked. The undulatus cloud variety tends to extend over more of the cloud surface. It has a more regular appearance of rows of ridges, rolls, or undulations. Asperitas is a cloud supplementary feature, which appears typically on just a patch or region of a cloud. It tends to have a well-defined base, and exhibits a much more chaotic and turbulent pattern of undulations—like a rough sea viewed from below. Is everyone paying attention at the back of the class?

THE LENGTH OF AN AIRCRAFT condensation trail, or contrail, gives an indication of the air conditions up at cruising altitude. When the air up there is dry, the water vapor produced by the aircraft doesn't form into droplets and no contrail appears. When it contains a lot more moisture, contrails form and freeze into ice crystals that then grow, splinter, and spread out in the high winds to form broad streaks of cloud extending right across the sky. When the air has a moisture content somewhere in between, the trail forms briefly like this before its droplets or ice crystals dissipate.

ABOVE A characteristically dramatic Altocumulus sunset floats above a soft bed of fog over Recanati, Italy, spotted by Marco Cingolani (Member 7,635).

OPPOSITE Contrail, spotted by Michael Warren (Member 37,489) from McNears Beach, Marin County, California, US.

THE POUCH-LIKE FEATURES known as mamma are usually noticed on the underside of storm clouds. This, after all, is where mamma look at their most dramatic, as in the example below, showing them hanging from the canopy that spreads out at the top of a Cumulonimbus cloud. But the same cloud lobes can also form under several other of the main cloud types. One example is Cirrus, shown right. Mamma hanging from the streaks of Cirrus clouds are far rarer and less easy to identify because they are much more subtle in appearance. Cirrus mamma is a classification for the cloudspotting connoisseur.

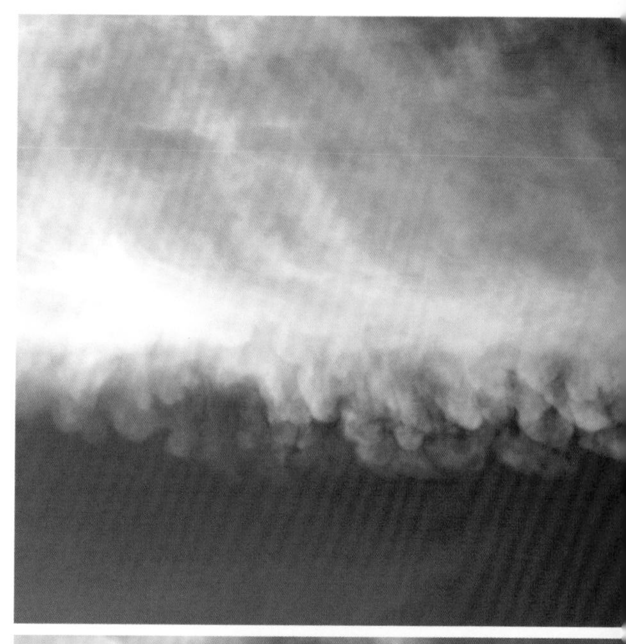

TOP Cirrus mamma, spotted by Kristof De Maeseneer (Member 32,680) over Lebbeke, Belgium.

RIGHT Cumulonimbus mamma, spotted by Alison Banks over Purcell, Colorado, US.

OPPOSITE *Le Nuage Rose, Antibes* (1916), by Paul Signac.

THE WARM, ROSY HUES that color clouds at the end of the day come from the journey sunlight makes to reach them. Take the cloud in Paul Signac's scene over Antibes, France. The Sun has dipped to the horizon, so its rays are passing sideways through the sky. They illuminate the mountain of cloud after a lengthy voyage through air, whose particles scatter them this way and that. The journey warms the colors of the sunlight because our atmosphere scatters blueish light a little better than reddish light. The cool blueish rays are teased out. The yellow- and pink-looking ones pass through. Signac's cloud has a golden summit since higher, less bulky atmosphere was traversed to reach it than the rosy base. Less still, the Cirrus wisps above. Even at this late hour, the high clouds shine bright white against the darkening sky.

So fine was the morning except for a streak of wind here and there that the sea and sky looked all one fabric, as if sails were stuck high up in the sky, or the clouds had dropped down into the sea.

From *To the Lighthouse* (1927), by Virginia Woolf.

ABOVE Cirrus reflections, spotted by Junichi Kai over Odo Yacht Harbour, Fukuoka City, Japan.

THE MOTHER-OF-PEARL colors of the stratospheric nacreous clouds make them one of the most beautiful formations. But there is a flip side to these pretty clouds. At altitudes of 10–25 kilometers (6–16 miles), they form up at the level of the ozone layer. And guess what? Their frozen particles serve as the perfect environments for the chemical reactions to take place in which the ultraviolet light-shading ozone is broken down by the CFC gasses we released into the atmosphere before their ban in the late 1980s. Damn.

ABOVE Nacreous or mother-of-pearl clouds, spotted over Kells, County Antrim, Northern Ireland, by Paul Bell (Member 5,248).

ABOVE We know pigs don't fly, but they do sometimes enjoy sitting on telephone cables. Spotted by Jeanette White over Everett, Washington, US.

OPPOSITE Twisting curls of Cirrus intortus dance above Stratus, weightless and free in the skies over Tasmania. Spotted by Susan McArthur over Burnie, Tasmania, Australia.

CLOUD ENTHUSIASTS TEND to have a bittersweet relationship with Stratus. Along with its cousins the higher Altostratus and the wetter Nimbostratus, this layer cloud is arguably responsible for the negative reaction most people have towards clouds in general. The misguided concept of "blue-sky thinking" was likely developed in response to a particularly gloomy Stratus formation. On the other hand, there is romance to be found in a low-lying Stratus mist. Alone with your thoughts in a landscape of successive shades of grey, you can lose yourself in deep contemplation. To gaze at Stratus is to confront the possibility that infinity exists. To feel like the last human alive.

CUMULUS MEDIOCRIS head to the shore of Lake Michigan. They're not scared of the water, they just form over the land as it heats up rapidly in the summer sunshine while the lake stays nice and cool. They might take a dip later, but only if they build into Cumulus congestus shower clouds and throw themselves in, piece by tiny piece.

ABOVE Summer Cumulus, spotted just north of Chicago, Illinois, US, by Paul Noah (Member 46,523).

OPPOSITE Stratus, spotted by Tom Keymeulen over Middelkerke, West Flanders, Belgium.

THIS PASTEL DRAWING by American artist Zaria Forman depicts rainfall onto a calm sea. The trails of showers are likely falling from Cumulus congestus towers whose bases are merged together into a continuous cloud cover. The trails are classed not as virga, which is when they fade in the sky as the precipitation evaporates away, but as praecipitatio, since they are reaching all the way to the surface.

ABOVE Detail from *Untitled No. 31* (2006), by Zaria Forman.

And what is Life?—An hour-glass on the run,
A mist retreating from the morning sun,
A busy, bustling, still repeated dream;
Its length?—A minute's pause, a moment's thought;
And happiness?—A bubble on the stream,
That in the act of seizing shrinks to nought.

From "What is Life?" (1820), by John Clare.

ABOVE Mist at sunrise over Loch Tummel, Perth, and Kinross, Scotland. Spotted with Altocumulus above by Ian Loxley (Member 1,868).

CLOUDS EXTENDING IN LONG ROWS like these are sometimes known as cloud streets. Since the tops of the clouds mark the top of thermal updrafts, and since they are aligned in the direction of the wind, such formations act like highways in the sky for gliders. The first reported glider flight using cloud streets was in 1935. It extended from Bayreuth in western Germany to Brno in the Czech Republic, a distance of over 500 kilometers (300 miles). Glider pilots have been on the lookout for them ever since.

LEFT Cloud streets, also known as Cumulus radiatus, spotted by Seth Adams on a flight over the Southern Appalachian Mountains, US.

WHEN A SHARP LINE APPEARS cut from a cloud like this, it has invariably been caused by an aircraft. It is an example of the cavum formation, also known as a fallstreak hole, when the very cold droplets in a cloud layer start to freeze and fall below as ice crystals. If these dissipate away in warmer, drier air beneath the cloud, all that is left is a hole—or in this case, a line. An aircraft can trigger this freezing process by the cooling that happens within its wing vortices or by the tiny particles in its exhaust acting as freezing nuclei. Such nuclei, whether they occur naturally as particles of dust, ash, or plant material or they are introduced artificially like this, are generally needed before cloud droplets can freeze. A man-made cavum like this is known as a dissipation trail, or distrail.

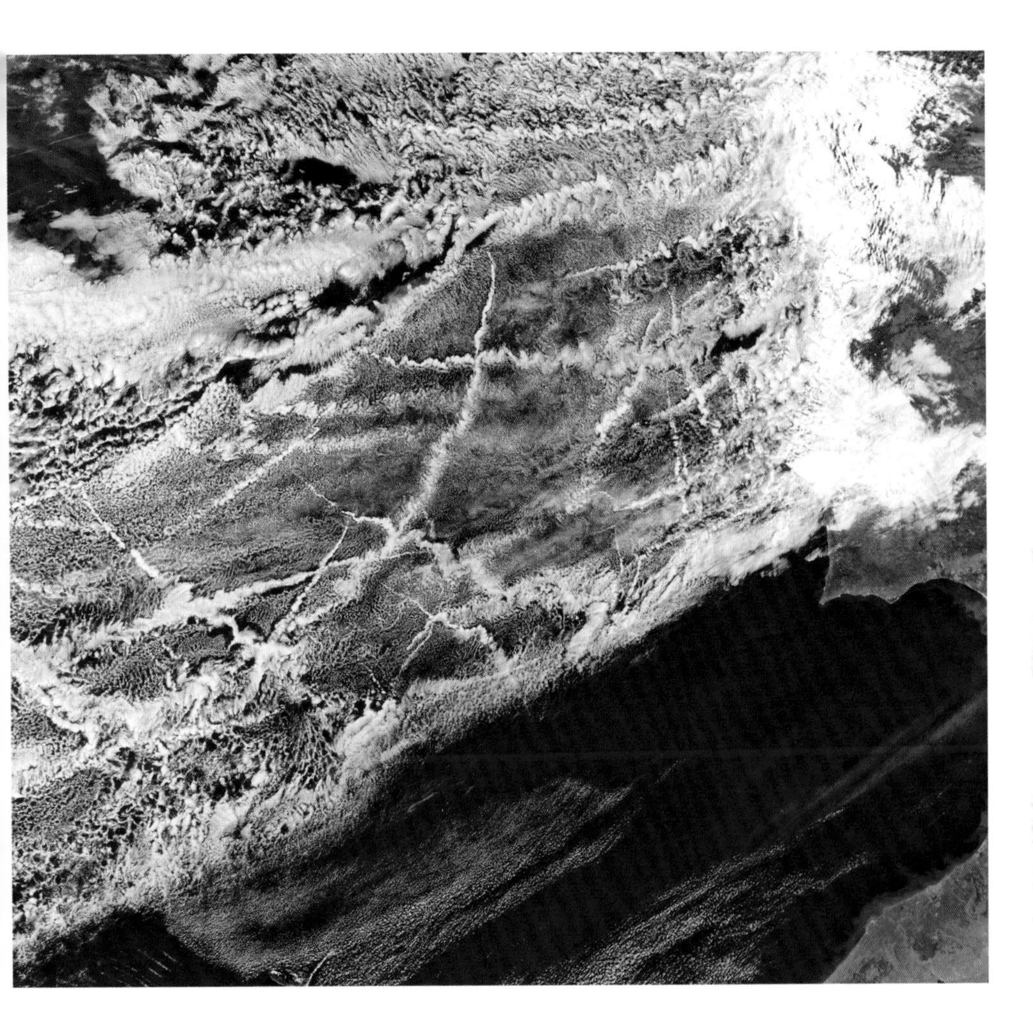

WE ARE USED TO SEEING lines of cloud, known as condensation trails, or contrails, form behind aircraft. But the cloud tracks shown here wandering across the ocean off the coast of Portugal and Spain are formed by ships. They are known as ship tracks and they appear as water droplets form onto the tiny particles of pollution emitted in the ships' exhausts, which act as the nuclei onto which the water molecules start to gather. When atmospheric conditions are right for them to appear, ship tracks can, like these, extend hundreds of kilometers.

ABOVE Ship tracks, spotted by NASA's Aqua satellite.

OPPOSITE Two aircraft-induced cavum, known as distrails, spotted by Liam Greany over New Southgate, Enfield, England.

ANYONE FAMILIAR with the work of the Belgian surrealist painter René Magritte will have noticed that he was a cloudspotter. But he only ever seemed to spot one type of cloud: Cumulus. In this surreal juxtaposition of night and day, called *The Empire of Light, II*, Magritte's Cumulus are flat, fair-weather ones. These smallest species of Cumulus re known as humilis.

ABOVE Detail from *The Empire of Light, II* (1950), by René Magritte.

OPPOSITE A banner cloud spotted by James Williams over the Jungfrau, Switzerland, showing iridescent colors and "turbulence holes" (our name for them, since they've yet to be given an official classification).

IN THE UNTAMED, TURBULENT AIR downwind of mountains, fine cloud filigrees can develop. A prominent peak like the Jungfrau in the Swiss Alps often produces formations known as banner clouds, which stretch out in the lee of the mountain. These clouds can exhibit two distinctive characteristics: bands of pastel colors known as cloud iridescence, and holes that look as if they have been punched out of them. The colors are caused by the bending, or diffracting, of light around the banner cloud's tiny droplets. The holes are turbulence made visible. Where strong winds form powerful eddies in the lee of the peak, vortices suck down pockets of air and tear holes in the cloud. When the atmospheric conditions are just right, the fierce churn of mountain winds produces one of the most delicate, lacy cloud formations of all.

Thousands have lived without love, not one without water.

From "First Things First" (1956), by W. H. Auden.

ABOVE Double rainbow, spotted over Bukovník, Plzeň, Czech Republic, by Karel Jezek (Member 34,987).

LENTICULARIS, THE SMOOTHEST OF CLOUDS, make Cumulus look a bit scruffy. The two are produced in very different ways. Cumulus bubble up on top of thermals. Lenticularis emerge within the smooth airflow downwind of hills and mountains. How nice to see them getting along.

ABOVE Altocumulus lenticularis and Cumulus, spotted on a flight over Navajo, Arizona, US, by Tom Bean (Member 41,135).

That which is below is like that which is above and that which is above is like that which is below.

From the *Emerald Tablet*, translated by Sir Isaac Newton (c. 1680).

ABOVE Cirrostratus, spotted at sunrise over Lake Manyara, Tanzania, by Abbas Virji (Member 39,576).

OPPOSITE Every majestic view is improved with the presence of clouds. Here, Switzerland's Eiger mountain floats between a layer of Stratus in the Grindelwald valley below and a layer of Altocumulus in the sky above. Spotted by John Callender (Member 26,942).

MOST PEOPLE THINK OF STRATUS as a low, extended blanket of cloud that fills valleys, covers the sea, and shrouds tall buildings. But it does also have a far less smothering incarnation, known as Stratus fractus. This manifests as delicate, transient shreds, which are usually found hugging the sides of hills or mountains. Such scraps of cloud can form when saturated air cools as it gently lifts up an incline. The fractus species of Stratus is not only subtler than the usual overcast form, known as nebulosus, but it is also more evocative. It adds a sense of ancient mystery to a landscape.

ABOVE Stratus fractus, spotted by Gillian Edkins (Member 42,894) over Westfjords, Iceland.

LANDSCAPE PAINTERS are attuned to the moods of the atmosphere more than most people. "It is the sky that makes the earth so lovely at sunrise, and so splendid at sunset," wrote Thomas Cole, the 19th-century artist who founded the Hudson River School. "In the one it breathes over the earth the crystal-like ether, in the other liquid gold." While Jon Schueler, a 20th-century American painter who worked in Mallaig, on the west coast of Scotland, wrote: "It is most particularly the brooding, storm-ridden sky over the Sound of Sleat in which I find the living image of past dreams . . . Here I can see the drama of nature charged and compressed. Lands form, seas disappear, worlds fragment, colors merge or give birth to burning shapes . . ."

TOP Delicate Stratus fractus grace the slopes of the Catskill Mountains, New England, US, in *Sunny Morning on the Hudson River* (1827), by Thomas Cole.

LEFT The stormy view out across the Sound of Sleat from Mallaig, Scotland, in *Light and Black Shadow* (1977), by Jon Schueler.

I can be jubilant one moment and pensive the next, and a cloud could go by and make that happen.

Bob Dylan, interviewed in 1997 by the *New York Times*.

ABOVE Cumulus fractus going by, spotted from the flight deck by John Gale (Member 15,702).

OPPOSITE Halo phenomena, spotted by Kevin Schafer (Member 46,954) over Iceland.

IN THIS SKY OF MIXED CLOUDS over Iceland, two arcs of light have formed around the Sun as it shines through the edge of a thick patch of Cirrus. The innermost arc is part of the common, circular 22-degree halo. The outer arc is more rare. It is a circumscribed halo, and is unusual among halo phenomena because its shape varies greatly with the Sun's elevation. When the Sun is very high, this halo appears as a simple circle too. But as the Sun sinks a little lower, like it has in this image, the shape of the circumscribed halo becomes more oval. Lower still, and the sides of the oval droop, making it more of a kidney shape. By the time the angle between Sun and horizon gets down to 30 degrees the optical effect has split, appearing now as two tangent arcs, one above and one below the Sun. Say all this next time someone looks up and asks, "What's with that weird double rainbow?"

ABOVE Two ships sail at sunset across a windswept sea. No, wait a minute. Actually, it's a rolling blanket of fog over the Boston skyline, spotted from the flight deck by Peter Leenen (Member 32,762).

OPPOSITE A 22-degree halo, sun dog, and lower tangent arc caused by diamond dust, spotted by David Malpas over Summit Camp, Østgrønland, Greenland.

He who binds to himself a joy
Does the winged life destroy;
But he who kisses the joy as it flies
Lives in eternity's sun rise.

From *The Notebook of William Blake* (c. 1793).

I'll tell you how the sun rose, –
A ribbon at a time.

The steeples swam in amethyst,
The news like squirrels ran.

The hills untied their bonnets,
The bobolinks begun.

Then I said softly to myself,
"That must have been the sun!"

From "I'll Tell You How the Sun Rose" (1861), by Emily Dickinson.

ABOVE Sunrise of Altocumulus and Altostratus, spotted over Finnmark, Norway, by Jon Hearn (Member 32,683).

THE IMAGES OF CLOUDS we share through the Cloud Appreciation Society, like those you'll find in reference books, are chosen to look dramatic. Everyone favors examples where the formation dominates the whole sky. And while this might help to show the classification's distinguishing features, it can engender a sense of cloudspotting inadequacy. Do you yearn for distant lands where the clouds are more spectacular, more exotic? This is to miss the point of cloudspotting. It is a frame of mind that finds the exotic in the everyday. Take the fluctus cloud, which looks like a series of curling ocean breakers. The formation is always described as rare, but in reality only the dramatic examples are. You don't need to spot one dominating the whole sky. A subtle example like this is just as valuable—perhaps more so, because you were the only one to notice it.

ABOVE Fluctus forming on a Cumulus humilis cloud, spotted over Aarhus, Central Jutland, Denmark, by Søren Hauge (Member 33,981).

Picture Credits

2 Pat Cooper; 15 Stephen Ingram; 16 *Cloudy Mountains* (before 1200) by Mi Youren. In the collection of the Metropolitan Museum of Art, New York, US. Ex. coll: C. C. Wang Family, Purchase. Gift of J. Pierpont Morgan, by exchange, 1973. Photograph: age fotostock/Alamy Stock Photo; 17 Ian Boyd Young; 18 Danielle Malone; 19 Sarah Jameson; 20 Keith Edmunds; 21 Frank Leferink; 22 Debbie Whatt; 23 Søren Hauge; 24 NASA/JPL-Caltech/Space Science Institute; 25 *New Mexico Recollection* (1922–23) by Marsden Hartley. In the collection of the Blanton Museum of Art, Austin, Texas, US. Photograph: Historic Images/Alamy Stock Photo; 26 NASA/JSC; 28 Rauwerd Roosen; 29 NASA image courtesy Jeff Schmaltz, LANCE/EOSDIS MODIS Rapid Response Team at NASA GSFC; 30 Rodney Jones; 31 Poppy Jenkinson; 32 Cloud studies (1820–22) by John Constable, in the collection of the Yale Center for British Art, New Haven, Connecticut, US. Photograph: Mark Richardson; 33 Elise Bloustein; 34 Katalin Vancsura; 35 (top) Ram Broekaert; 35 (bottom) Richard Ghorbal; 36 Ebony Willson; 37 NASA/JPL/University of Arizona—HiRISE; 38 *Seashore by Moonlight* (c. 1835–36) by Caspar David Friedrich. In the collection of the Hamburger Kunsthalle, Hamburg, Germany. Photograph: Art Collection 4/Alamy Stock Photo; 39 Kelly Hamilton/Gingham Sky Photography; 40 Sarah Nicholson; 41 NASA/JPL-Caltech/SwRI/MSSS/Bjorn Jonsson; 42 David Law; 43 Bibliothèque de Genève, Switzerland; 44 (top) Greg Dowson; 44 (bottom) *Starry Night* (1889) by Vincent van Gogh. In the collection of the Museum of Modern Art, New York, US. Photograph: FineArt/Alamy Stock Photo; 45 Althea Pearson; 46 Christopher Watson; 47 J.P. Harrington and K.J. Borkowski (University of Maryland) and NASA/ESA; 48 NASA/MODIS; 49 Hetta Gouse; 50 Gavin Pretor-Pinney; 51 Mike Brown; 52 Rural landscape with later cloud studies by Luke Howard (1772–1864) c. 1808–11 (w/c on paper), Kennion, Edward (1744–1809)/Science Museum, London, UK/Bridgeman Images; 53 Sim Richardson; 54 ESA/NASA; 55 Katrina Whelen; 56 *Ground Swell* (1939) by Edward Hopper . On display at the National Gallery of Art, Washington DC, US, Corcoran Collection (Museum Purchase, William A. Clark Fund). © Heirs of Josephine Hopper/Licensed by Artists Rights Society (ARS) NY/DACS, London 2019; 57 Debra Ceravolo; 58 Karin Enser; 59 (top) Michael Sharp; 59 (bottom) ESA/NASA-A. Gerst; 60 *View of Bievre-sur-Gentilly* (c. 1895) by Henri Rousseau. Private collection. Photograph: The Picture Art Collection/Alamy Stock Photo; 61 Chris Devonport; 62 Jeff Schmaltz, MODIS Land Rapid Response Team, NASA GSFC; 63 Joan Laurino; 64 Christine Alico; 65 Michael Schwartz; 66 (left) Badr Alsayed; 66 (right) Alain Buysse; 67 Matthew Curley; 68 *Above the Clouds I* (1962/1963) by Georgia O'Keeffe. O'Keeffe, Georgia (1887–1986): Above the Clouds 1, 1962–63. Santa Fe, The Georgia O'Keeffe Museum. Oil on canvas. 36 1/8 x 48 1/4 in. 1997.05.14. Gift of the Burnett Foundation and the Georgia O'Keeffe Foundation. Photo: Malcolm Varon 2001. © 2019 Photo Georgia O'Keeffe Museum, Santa Fe/Art Resource/Scala, Florence; 69 Frieder Wolfart; 70 Frank Povah; 71 Jelte Vredenbregt; 72 Tom Bean; 73 Maria Wheatley; 74 Matt Friedman; 75 *A Balloon Prospect from Above the Clouds* (1786) by Thomas Baldwin. Department of Special Collections, Memorial Library, University of Wisconsin-Madison, Madison, Wisconsin, US; 76 Julie Magyar Africk; 77 NASA; 78 Hans Stocker; 79 Terence Pang; 80 *Winter Scene in Moonlight* (1869) by Henry Farrer. From the Collection of the Metropolitan Museum of Art, New York, US. Purchase, Morris K. Jesup Fund, Martha and Barbara Fleischman, and Katherine and Frank Martucci Gifts, 1999; 81 (left) James Brooks; 81 (right) Jean-Christophe Benoist; 82 Adolfo Garcia Marin; 83 Christina Brookes; 84 ESA/NASA-A. Gerst; 85 Sylvia Fella; 86 Chito L. Aguilar; 87 Edward Hannen; 88 Norman Kuring, NASA's Ocean Biology Processing Group; 89 Tim de Wolf; 90 NASA/JPL/Texas A&M/Cornell; 91 Andy Sallee; 92 Melyssa Wright; 93 Tom Bean; 94 Melyssa Wright; 95 (left) Antonio Martin; 95 (right) Karel Jezek; 96 *Vädersolstavlan* (c. 1535) by Urban Målare (anonymous painter) or Jacob Elbfas. In Storkyrkan Cathedral, Stockholm, Sweden; 97 Kenneth R. Carden; 98 Maxine Hill; 99 Jente De Schepper; 100 NASA; 101 robertharding/Alamy Stock Photo; 102 *Clouds over Water: Moonlight, 1796* by J.M.W. Turner. D00828 by Tate Images/Digital Image © Tate, London 2014 *from* Studies near Brighton Sketchbook (Finberg XXX), Clouds over Water: Moonlight; 103 Patrick Dennis; 104–105 Nienke Lantman; 106 Daniel Fox; 107 Lana Cohen; 108 (top) NASA/ESA/STScl; 108 (bottom) NASA/JPL-Caltech/Space Science Institute; 109 Richard Corrigall; 110 Sallie Tisdale; 111 *South Wind, Clear Sky* (1830) by Katsushika Hokusai. From the collection of the Metropolitan Museum of Art, New York, US. Henry L. Phillips Collection, Bequest of Henry L. Phillips, 1939; 112 Jaap van den Biesen and Nienke Edelenbosch; 113 ESA/NASA; 114 Petr Kratochvil; 115 Filip Gavanski; 116 Jeremy Hanks; 117 (top, middle and bottom) Jean Louis Drye; 118 NASA; 119 Maarten Hoek; 120 Cecelia Cooke; 121 *Genesis, The Creation: Division of Sea and Earth* (c. 1467) by The Master of Feathery Clouds. © Koninklijke Bibliotheek, National Library of the Netherlands; 122 Ross McLaughlin; 123 Marcus Murphy; 124 Darya Light; 125 NASA Earth Observatory image by Jesse Allen, using data from the Land Atmosphere Near real-time Capability for EOS (LANCE); 126 JSC/NASA; 127 John Findlay; 128 Christina Connell; 129 Randolph Harris; 130 NOAA-NASA-GOES Project; 131 Linda Eve Diamond; 132 Jenny Shanahan; 133 Nicole Bates – Eight Skies; 134 *Buddha Traverse Le Gange*, an anonymous illustration from "Vie Illustrée du Bouddha Cakayamouni" in *Recherches sur les Superstitions en Chine* (1929), Vol 15, by Henri Doré. ; 135 (left) David Rudas; 135 (middle) Judy Shedd; 135 (right) James Morrison; 136 Graeme Blissett; 137 NASA/

ISS Crew Earth Observations Facility and Earth Science and Remote Sensing Unit, Johnson Space Center; 138 *The Village by the Lake* (1929) by Paul Henry. In the collection of Museums Sheffield, Sheffield, UK. Photo © Museums Sheffield. ; 139 Frits Kuitenbrouwer; 140 Peter Leenen; 141 Robyn Molnar; 142 NASA image by Jeff Schmaltz, LANCE/EOSDIS Rapid Response #nasagoddard; 143 Heather Prince; 144 Laura Simms; 145 National Oceanic & Atmospheric Administration (NOAA), NOAA Central Library. Published in the Monthly Weather Review, December 1902, from the US Weather Bureau; 146 Graham Billinghurst; 147 Tom Bean; 148 James Tromans; 149 Juergen K. Klimpke; 150 NASA; 151 Fiorella Iacono; 152 *Gray and Gold* (1942) by John Rogers Cox. *Gray and Gold*, 1942. John Rogers Cox (American, 1915–1990). Oil on canvas, framed: 116 x 152 x 12.5cm (45 11/16 x 59 (13/16) x 4 15/16 in.); unframed; 91.5 x 151.8cm (36 x 59 3/4in). The Cleveland Museum of Art, Mr. and Mrs. William H. Marlatt Fund 1943.60. ; 153 (left) Karel Jezek; 153 (right) Monica Nitteberg; 154 Jorge Figueroa Erazo; 155 Nizma Arifin; 156 Norman Kuring, NASA's Ocean Color web; 157 "Falling stars as observed from the balloon," an illustration from *Travels in the Air* (1871) by James Glaisher, Camille Flammarion, Wilfrid de Fonvielle and Gaston Tissandier. National Oceanic & Atmospheric Administration (NOAA), Treasures of the NOAA Library Collection; 158 Adam Littell; 159 Mark Hayden; 160 John Bigelow Taylor; 161 Joost van Ekeris; 162 (top) ESA/GCP/UPV/EHU Bilbao, CC BY-SA 3.0 IGO; 162 (bottom) Raymond Kenward; 163 Kym Druitt; 164 Illustration for Henry Wadsworth Longfellow's "The Rainy Day" by Myles Birket Foster. In the collection of the National Gallery of Art, Washington DC, US; 165 NASA; 166 (left) Chase Vessels; 166 (right) Leslie Cruz; 167 Dmitry Kolesnikov; 168 NASA Earth Observatory image by Joshua Stevens, using Landsat data from the US Geological Survey and topographic data from the Shuttle Radar Topography Mission; 169 Laura Stephens, www.amindfulmess.com; 170 James McAllister; 171 Christina Brookes; 172 Ko van Hespen; 173 *Equivalents* (1926) by Alfred Stieglitz. From the collection of the Metropolitan Museum of Art, New York, US. Alfred Stieglitz Collection, 1949; 174 Lauren Antanaitis; 175 "Cross-sections of large hailstones" illustration from *L'Atmosphere* by Camille Flammarion. National Oceanic & Atmospheric Administration (NOAA), Treasures of the NOAA Library Collection; 176 Lodewijk Delaere; 177 Dennis Olsen; 178 NASA/JPL-Caltech/Space Science Institute; 179 Fresco by Giotto, in the Basilica of St Francis in Assisi, Italy. Photograph: The Picture Art Collection/Alamy Stock Photo; 180 ESO/Y. Beletsky. Licensed under the Creative Commons Attribution 2.0 Generic (CC BY 2.0) license (https://creativecommons.org/licenses/by/2.0/legalcode) ; 181 Margot Redwood; 182 *Lightning* (1909) by Mikalojus Ciurlionis. In the collection of the M. K. Ciurlionis National Art Museum in Kaunas, Lithuania; 183 Chris Damant; 184 Michael Warren; 185 Roger Lewis; 186 Paula Maxwell; 187 (top) Michela Murano and Valeriano Perteghella; 187 (bottom) Sun dogs illustration from the *Nuremberg Chronicle*, 1493. Photograph: Science History Images/Alamy Stock Photo; 188 Steven Grueber; 189 ESA/Hubble, R. Sahai and NASA; 190 Suzanne Winckler; 191 *Wanderer Above the Sea of Fog* (1817) by Caspar David Friedrich. In the collection of the Hamburger Kunsthalle, Hamburg, Germany; 192 Dennis Paul Himes; 193 John Callender; 194 ISS Crew Earth Observations Facility and the Earth Science and Remote Sensing Unit, Johnson Space Center; 195 Azhy Chato Hasan; 196 Phil Chapman; 197 USPS; 199 Pete Herbert; 200 *Before the Storm* (1890) by Isaac Levitan. In the collection of the Smolensk State Museum Reserve, Smolensk, Russia; 201 Suzanne Winckler; 202 NASA; 203 Luda Sinclair; 204 Diagram of rainbow optics from *Discours de la Méthode* (1637) by René Descartes; 205 Tania Ritchie; 206 Justin Parsons; 207 (top) NASA; 207 (bottom) Fir0002/Flagstaffotos. Licensed under the Creative Commons Attribution-NonCommercial 3.0 Unported (CC BY-NC 3.0) license (https://creativecommons.org/licenses/by-nc/3.0/legalcode) ; 208 Charlie Gray; 209 Tiziano Bartolucci; 210 Fernando Flores; 211 © Diller, Scofidio + Renfro; 212 *Landscape with a View of the Valkhof, Nijmegen* by Aelbert Cuyp. National Galleries of Scotland. Purchased with the aid of the Art Fund 1972 (in recognition of the services of the Earl of Crawford and Balcarres to the Art Fund and the National Galleries of Scotland); 213 Graham Billinghurst; 214 Matt Minshall; 215 Dave Hall; 216 Martin Foster; 217 *The Olive Trees* by Vincent van Gogh. In the collection of the Museum of Modern Art, New York, US. Mrs. John Hay Whitney Bequest ; 218 "Clouds" from *Orbis Pictus* 1658) by John Amos Comenius; 219 Tania Ritchie; 220 Harriet Aston; 221 Marjorie Perrissin-Fabert; 222 NASA; 223 Patrick Dennis; 224 (top) NASA; 224 (bottom) Judy Taylor; 225 Richard Ghorbal; 226 NASA/JPL-Caltech/SwRI/MSSS/Gerald Eichstädt/Seán Doran; 227 Thorleif Rødland; 228 *View of Toledo* by El Greco. In the collection of the Metropolitan Museum of Art, New York, US. H. O. Havemeyer Collection, Bequest of Mrs. H. O. Havemeyer, 1929; 229 Jelte van Oostveen; 230 *The Great Wave, Sète* (1857) by Gustave Le Gray. In the collection of the Metropolitan Museum of Art, New York, US. Gift of John Goldsmith Phillips, 1976; 231 David Rosen; 232 Gary Davis; 233 Paul Jones; 234 (top left) Fred Ohlerking; 234 (top right) Graham Blackett; 234 (bottom left) Lauren Antanaitis; 234 (bottom right) Anne Downie; 235 (top left) Hélène Condie; 235 (top right) Saskia van der Sluis; 235 (bottom left) Sugata Kuila; 235 (bottom right) Peter Beuret; 236 NASA/JPL-Caltech/Space Science Institute; 237 Carole Pereira; 238 *Study of Clouds with a Sunset near Rome* by Simon Denis. In the collection of the J. Paul Getty Museum, Los Angeles, US; 239 Beth Holt; 240 Colleen Thomas; 242 Juho Holmi; 243 Vicki Kendrick; 244 Easton Vance; 245 *The British Channel Seen from the Dorsetshire Cliffs* (1871) by John Brett. N01902 by Tate Images/Digital Image © Tate, London 2014; 246 NASA/STScl Digitized Sky Survey/Noel Carboni; 247 Ross Hofmeyr; 248 Iridescent clouds—looking north from the Ramp on Cape Evans, Aug 9, 1911 (w/c on paper), Wilson, Edward Adrian (1872–1912)/Scott Polar Research Institute, University of Cambridge, UK / Bridgeman Images; 249 Hallie Rugheimer; 250 Kristina Machanic; 251 Marie Dent; 252 (top) NASA/JPL; 252 (bottom) Althea Pearson; 253 *Bleaching Ground in the Countryside Near Haarlem* (1670) by Jacob van Ruisdael. Bleaching Ground in the Countryside near Haarlem, 1670 (oil on canvas), Ruisdael, Jacob Isaaksz. or Isaacksz. van (1628/9–82) / Kunsthaus, Zurich, Switzerland / Bridgeman Images; 254 Baiyan Huang; 255 Carlyle Calvin; 256–257 *Nimbus Dumont, 2014* by Berndnaut Smilde. Courtesy of the artist and Ronchini Gallery; 258 "Frigga Spinning the Clouds" by J. C. Dollman, from *Myths of the Norsemen* (1922) by H. A. Guerber; 259 Lilian van Hove; 260 Hannah Hartke; 261 Mike Cullen; 262 *Winter Landscape 2* by Alex Katz. In the collection of the High Museum of Art, Atlanta, Georgia, US. © Alex Katz/VAGA at ARS, NY and DACS, London 2019; 263 Tom Montemayor/McDonald Observatory; 264 Wayde Margetts; 265 James Helmericks; 266 (top) Brett King; 266 (bottom) Gavin Pretor-Pinney; 268 Paul Martini; 269 NASA/Mark Vande Hei; 270 Fiona Graeme-Cook; 271 Detail from frontispiece of *An Invective Against Cathedral Churches, Church-Steeples, Bells, etc* (1656) by Samuel Chidley. British Library/Bridgeman Images; 272 NASA/ISS; 273 Nicole Bates; 274 (Patricia) Keelin; 275 Enrique Roldán; 276 Daisy Dawson; 277 Elizabeth Freihaut; 278 ESA/Hubble, NASA, A. Simon (GSFC) and the OPAL Team, J. DePasquale (STScl), L. Lamy (Observatoire de Paris) ; 279 Matteo Pessini/Alamy Stock Photo; 280 Henrik Välimäki; 281 Peter van de Bult; 282 Anne Hatton; 283 Margaret D. Webster; 284 Sofie Bonte; 286 Paul Martini; 287 Doug Short; 288 *Roof Ridge of Frederiksborg Castle with View of Lake, Town and Forest* (1833–34) by Christian Købke. In the collection of the Statens Museum for Kunst in Copenhagen, Denmark; 289 Emily Watson; 290 Stephen Ingram; 291 Maria Lyle; 292 Roberval Santos; 293 Busra Karademir; 294 Jean Gray; 295 Stephanie Arena; 296 (top) 24 NASA/JPL-Caltech/ Space Science Institute; 296 (bottom) Fiona Semmens; 297 Marty Bell; 298 Patty Kjobmand Cashman; 299 *The Translation of the Holy House of Loreto* (mid-1490s), attributed to Saturnino Gatti. In the collection of the Metropolitan Museum of Art, New York, US. Gwynne Andrews Fund, 1973; 300 Peter Dayson; 301 Marc van Workum; 302 Elizabeth Watson; 303 Tony Hoffman; 304 Shotsy Faust; 305 Anthony Skellern; 306 *Clouds Over the Black Sea* (1906) by Boris Anisfeld. Boris Anisfeld (Russian, 1879–1973). Clouds over the Black Sea–Crimea, 1906. Oil on canvas, 491/2 x 56in (125.7 x 142.2cm). Brooklyn Museum, gift of Boris Anisfield in memory of his wife, 33.416. Photo: Brooklyn Museum; 307 Nienke Lantman; 308 *Clouds and Sunbeams Over the Windberg Near Dresden* (1857) by Johan Christian Dahl. In the collection of the National Gallery of Norway; 309 (left and right) Mary Stivison; 310 Laura Simms; 311 *Burial of the Sacred Wood* by Piero della Francesca. Church of San Francesco, Arrezo, Italy; 312 Jeff Schmaltz, MODIS Rapid Response Team, NASA/GSFC; 313 Sinead Hurley; 314 Lucy Goldner; 315 Søren Hauge; 316 Eystein Mack Alnaes; 317 (left) George Preoteasa; 317 (middle) Jan McIntyre; 317 (right) Thibaut de Jaegher; 318 Roberval Santos; 319 Ross McLaughlin; 320 Mural by Howard Crosslen at the National Center for Atmospheric Research, Boulder, Colorado, US. Photo: Gavin Pretor-Pinney; 321 ESA/NASA; 322 Jammin Palmer; 323 Renee Gerber; 324 Jeff Schmaltz, MODIS Rapid Response at NASA GSFC; 326 Deborah Milics; 327 Paul Harwood; 328 *Mishima Pass in Kai Province* (around 1830) by Katsushika Hokusai. In the collection of the Metropolitan Museum of Art, New York, US. Rogers Fund, 1914; 329 Gary McArthur; 330 Michael Warren; 331 Marco Cingolani; 332 (top) Kristof De Maeseneer; 332 (bottom) Alison Banks; 333 *Antibes (La Pinède)* by Paul Signac. Private collection; 334 Junichi Kai; 335 Paul Bell; 336 Jeanette White; 337 Susan McArthur; 338 Tom Keymeulen; 339 Paul Noah; 340 *Untitled No. 31* (2006) by Zaria Forman. Courtesy of the artist Zaria Forman; 341 Ian Loxley; 342 Seth Adams; 344 Liam Greany; 345 NASA/Jeff Schmaltz, LANCE/EOSDIS Rapid Response; 346 *The Empire of Light, II* by René Magritte. In the collection of the Museum of Modern Art, New York, US. © ADAGP, Paris and DACS, London 2019. ; 347 James Williams; 348 Karel Jezek; 349 Tom Bean; 350 John Callender; 351 Abbas Virji; 352 Gillian Edkins; 353 (top) *Sunny Morning on the Hudson River* (1827) by Thomas Cole. In the collection of the de Young Museum, San Francisco, US; 353 (bottom) *Light and Black Shadow* (1977) by Jon Schueler. Jon Schueler (1916–1992), *Light and Black Shadow*, 1977, 69 x 76in/175.25 x 193cm, oil on canvas (o/c 876). © Jon Schueler Estate; 354 John Gale; 355 Kevin Schafer; 356 Peter Leenen; 357 David Malpas; 358 Jon Hearn; 359 Søren Hauge

42 "Fog" from THE COMPLETE POEMS OF CARL SANDBURG. Copyright © 1969, 1967 by Lillian Steichen Sandburg. Printed by permission of Houghton Mifflin Harcourt Publishing Company. All rights reserved; 67 From *The Wisdom of Water* by John Archer (Allen & Unwin, 2008). Reproduced with permission of Allen & Unwin Pty Ltd; 139 printed with the permission of Coleman Barks; 274 From THE BOOK OF MERLYN: THE UNPUBLISHED CONCLUSION TO "THE ONCE AND FUTURE KING" by T.H. White, Copyright © 1977. By permission of the University of Texas Press; 323 From EIGHT LITTLE PIGGIES: REFLECTIONS IN NATURAL HISTORY by Stephen Jay Gould. Copyright © 1993 by Stephen Jay Gould. Used by permission of W. W. Norton & Company, Inc.; 353 Extract from *The Sky* by Jon Schueler, Whitney Museum of American Art, brochure *Jon Schueler*, April 24–May 25, 1975.

INDEX